MATTER AND MOTION

RICHARD P. JACOBS   B.S. from the University of Wisconsin, Milwaukee with majors in Physics and Education; M.S. from the University of Wisconsin, Milwaukee with a major in the teaching of Physics. NSF grant at Montana State University in nuclear and instrumental chemistry. A Physical Science Instructor, the Science Department Chairman, and the Curriculum Coordinator at Grafton High School, Grafton, Wisconsin.

MATTER AND MOTION

PHYSICAL SCIENCE

RICHARD P. JACOBS

Science Department Chairman, Physical Science Instructor,
and Curriculum Coordinator at
Grafton High School
Grafton, Wisconsin

Cambridge Book Company
488 Madison Avenue, New York, N.Y. 10022

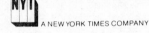
A NEW YORK TIMES COMPANY

Photograph credits:
Front cover: NASA. Unit 1: Woods Hole Oceanographic Institute/Sybil Shelton from Monkmeyer. Unit 2: Graphic Arts, University of California/Bethlehem Steel Corporation/Lawrence Radiation Laboratory, Berkeley/Bettman Archives/ Granger Collection. Unit 3: *The New York Times*/Bell Labs. Unit 4: UPI/Western Electric/New York Public Library. Unit 5: Bell Labs/IBM Corporation. Unit 6: Ringling Bros. Barnum & Bailey Combined Shows, Inc./Editorial Photocolor Archives/Helio Associates.

Cover Design: Lenni Schur
Text Design: Lenni Schur
Layout: Lenni Schur, Walter Schwarz
Artwork: Felix Cooper, Walter Schwarz, Barnett Graphics

Library of Congress Catalog Number: 75-273-75

CONTENTS

MATTER AND MOTION

PROPERTIES OF MATTER

Matter, as you know, exists in three states: solid, liquid, and gaseous. In this unit you will investigate many other physical properties of matter. The unit opening photograph shows a mechanical claw operated from *Alvin*, an American deep-diving craft, taking a sample of lava from the floor of the mid-Atlantic rift valley. The sampling was but one of many taken during project FAMOUS, which stands for French-American Mid-Ocean Undersea Study. Can you explain why the hot liquid lava bubbling up from the rift becomes solid in the ocean water?

YOUR OBJECTIVE: To learn how to find the length and volume of quantities of solids and liquids using metric units of measurement.

One characteristic or property of matter is that it takes up space. The amount of space that an object takes up is called its volume. Before the volume of an object can be found it is necessary to know certain measurements of the object. For instance, the volume of some objects can be found by measuring the length, width and height, and multiplying the three measurements (FIG. 1-1). A short way of expressing this calculation is by the equation $V = l \times w \times h$. Measurements such as length, width and height are called linear measurements because they are measurements made along a *line*.

Some of the units for making linear measurements in the **English system** are the inch, foot, yard and mile. Multiplying, subtracting, adding and dividing the numbers you get when you measure in units of the English system can be difficult — especially if you are not skilled in arithmetic.

Twelve inches make up *one* foot. *Three* feet make up *one* yard. *Five thousand, two hundred and eighty* feet make up *one* mile. Because the relationship between units varies, it is not easy to calculate such things as the number of inches in a certain number of yards, the number of yards in a certain number of miles, or the number of inches in a mile. (TABLE 1-1).

TABLE 1-1 English System of Linear Units		
12 inches	=	1 foot
3 feet	=	1 yard
5,280 feet	=	1 mile

There is a much easier system to work with than the English system of measurement. It is called the *metric system*. In the metric system the standard unit of length is the meter, abbreviated m. The meter is about three inches longer than one yard (FIG. 1-2). All other metric units of length

FIGURE 1-1

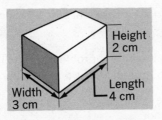

Height 2 cm
Width 3 cm
Length 4 cm

FIGURE 1-2 Which unit is larger — the meter or the foot?

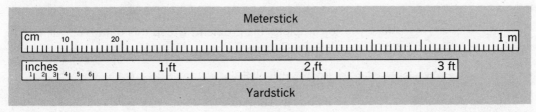

Meterstick
cm 10 20 1 m
inches 1 2 3 4 5 6 1 ft 2 ft 3 ft
Yardstick

TABLE 1-2
Metric System of Linear Units

100 cm	= 1 meter
1,000 mm	= 1 meter
1 km	= 1,000 meters

can be found by multiplying or dividing the number of meters by *ten or powers of ten*, such as *one hundred* or *one thousand* (TABLE 1-2).

Prefixes are added to the word *meter* to form other linear units of length. The prefix *milli* means 1/1,000. **The term millimeter means 1/1,000th of a meter. So, 1,000 millimeters equals one meter. A millimeter is abbreviated mm.** You can get an idea of the size of a millimeter if

Prefix Something added to the beginning of a word that changes its meaning.

you remember that it is about the same size as the thickness of the lead in a pencil.

The prefix *centi* means the 1/100 part of some unit. **A centimeter is 1/100th of a meter. So, 100 centimeters is equal to one meter. Centimeter is abbreviated cm.** The diameter of most pencils is a little smaller than one centimeter.

The prefix *kilo* means 1,000 times some unit. **There are 1,000 meters in 1 kilometer. One kilometer is about 5/8th of a mile. Kilometer is abbreviated km.** Once you learn the relationship within the metric system it becomes very easy to convert from one unit to another (TABLE 1-3).

TABLE 1-3

10 mm	= 1 centimeter
100 cm	= 1 meter
1,000 m	= 1 kilometer

BE CURIOUS 1-1: **Find out how many inches there are in 16 miles.**

Part A:

Watch or clock

Note the time that you begin this problem. From TABLE 1-1 select the figures you need to multiply to find inches per mile.

```
        feet per mile
  × _____ inches per foot
        inches per mile
```

Part B

Use your answer from Part A to compute Part B. How long did it take you to complete this problem? (Keep this data. You will need it to complete *Be Curious 1-2.*)

```
        inches per mile
  × _____ 16 miles
        inches in 16 miles
```

The volume of a rectangular solid can be found by multiplying the length by the width by the height (FIG. 1-1).

$$V = l \times w \times h$$

$$V = 4 \text{ cm} \times 3 \text{ cm} \times 2 \text{ cm} = 24 \text{ cm}^3$$

Some of the units used in measuring the volume of solids in the English system are cubic inches (**in.**^{3}) and cubic feet (**ft.**^{3}). The units of volume in the metric system

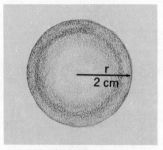

FIGURE 1-3 How would you figure the volume of this sphere?

include cubic centimeters (**cm**^{3}) and cubic meter (**m**^{3}). Often a *cubic centimeter* is abbreviated *cc*.

The volume of a sphere can be found from the equation

$$V = \frac{4}{3} \pi r^3.$$

The symbol π is equal to 3.14 and r is the radius of the sphere (FIG. 1-3). You can find equations for many other regular solids by consulting almost any mathematics text.

Cubic Units Units used in linear measurement to denote volume: Cubic means all *three* dimensions of the solid (l, w, h) are used to obtain the volume.
Cubic centimeter A cube 1 cm by 1 cm by 1 cm. Volume expressed in cubic centimeters tells how many 1 × 1 × 1 cm cubes would make up the solid.

BE CURIOUS 1-2: **Find out how many centimeters there are in 23 kilometers.**

Part A
Note the time you begin this problem. From TABLE 1-3 select the figures you need to multiply to find centimeters per kilometer.

Watch or clock

       meters per km
× _____ cm per meter
       cm per km

Part B
Use your answer to Part A to compute Part B. How long did it take you to complete this problem? Compare this time with the time taken to solve Part B in *Be Curious 1-1*. Which problem was easier to solve? Why?

       cm per km
× _____ 23 km
       cm in 23 km

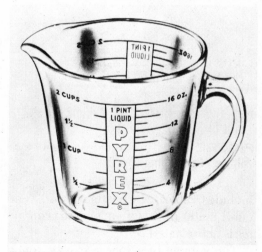

FIGURE 1-4

Finding the volume of a liquid is usually simple. The liquid is poured into a container that has graduations (measurement lines marked on its side.)

This measuring cup measures the volume of liquids in units of the English system (Fig. 1-4). The units commonly used in this system are the ounce, cup, pint, quart and gallon. Units of volume such as

cubic inches or cubic feet can also be used.

The liter is a unit of volume used in measuring liquids in the metric system. A liter is a volume of 1,000 cubic centimeters (Fig. 1-5). A liter is slightly larger than a quart.

Since the prefix milli means 1/1,000, 1,000 milliliters equals one liter. A milliliter (ml) is equal to one cubic centimeter since there are also 1,000 cubic centimeters in one liter (Fig. 1-6). Units of volume such as cubic millimeters or cubic meters are also used to measure the volume of liquids.

What if you wished to find the volume of an irregular solid object (Fig. 1-7a)? There is no mathematical equation that can be used to calculate its volume. Since the object is not a liquid, a graduated cylinder seems useless. However, a grad-

FIGURE 1-6 A quarter-liter measuring cup holds 250 milliliters (ml) or 2.5 deciliters (dl). What does the prefix *deci* mean? How many deciliters are in one liter? *(Sybil Shelton from Monkmeyer)*

FIGURE 1-5

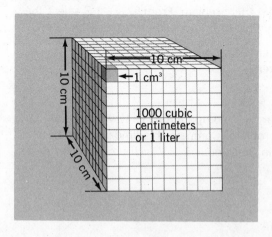

Find the volume, in metric units, of several different objects.

Marble
Red liquid, blue liquid
Wooden block, metal
block
Piece of clay washer
Metric ruler
Graduated cylinder
Water
Wash water, paper
towels

Measure the volumes of the objects listed in this investigation. Make all linear measurements to the nearest tenth of a centimeter. The volume of all solids should be stated in cubic centimeters, and those of liquids in milliliters. If there is more than one way to determine the volume of an object listed, do it each way and compare your results.

Roll the clay into a ball and find its volume using the equation

$$V = \frac{4}{3}\pi r^3$$

Shape the same amount of clay into a cube and find its volume from that shape. Now make it into an irregular shape and find its volume in this form. Explain your results.

Remember to clean and dry the cylinder and the objects dropped into liquid.

uated cylinder can be used indirectly. This is how it can be done. Pour water into a graduated cylinder to the 100-milliliter mark (Fig. 1-7b). Then drop the irregular solid into the cylinder. What happens to the water level (Fig. 1-7c)? It rises. The volume of the solid is equal to the difference in the level of the water before and after the object was dropped into the water. What is this volume?

In the example used, the volume of the object is 20 milliliters. Since one milliliter is equal to one cubic centimeter, you could also say that the volume is 20 cubic centimeters. Why was it suggested that you fill the graduated cylinder to the 100-ml mark before dropping the object into it? Could you have started with some other water level? Could you have started with 10 ml? Explain.

FIGURE 1-7 (a) Irregular solid object. (b) Graduated cylinder containing liquid. (c) Graduated cylinder containing liquid *and* the irregular solid object.

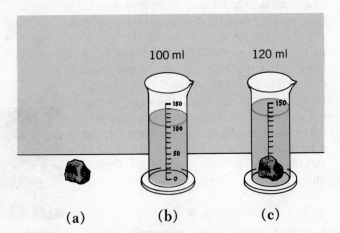

(a) (b) (c)

Most nations of the world use the metric system as their official national system of measurement. Even England has abandoned the English system in favor of the more easily used metric system. Those nations that have not adopted the metric system, including the United States, are seriously moving toward its adoption.

Changing the system of measurement used by a nation is not a simple task. Changing to the metric system will affect the lives of everyone in this country. The nation's football teams would probably find themselves with second down and eight meters to go on their opponents' fifteen meter line. When buying clothes, you would find that your waist measurement would be stated in the metric system. A waist size of 30 inches would change to one of 76 centimeters.

The changeover to metric in the United States would require considerable retooling. The gears, screw threads, axles and other parts of the industrial machines in this country are sized and constructed on the English system. A changeover to the metric system would require the gradual replacement of many of these machines and machine parts. Yardsticks, bathroom scales, milk bottles, and other measuring devices would have to be thrown away or changed to fit the metric system. Despite these problems other nations of the world have changed to the metric system. With all of the other industrial nations of the world changing to the metric system can you give economic reasons why the United States should also change? What is the width of the paper shown above in metric units? in English units?

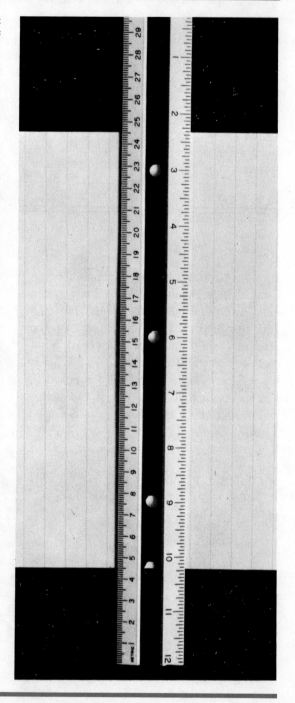

BE CURIOUS 1-4:

Small bag of sand
Small box
Graduated cylinder
Water

Find out if the volume of a certain amount of sand and of water equals the sum of the two separate volumes.

Pour sand into a graduated cylinder until the cylinder is a little less than half full. Record the volume of the sand. Pour the sand from the cylinder into a small box. Pour water into the graduated cylinder so that it is a little less than half full. Record the volume of the water. Then, add the sand to the water in the cylinder. Record the *total* volume of the sand and water. How does this compare to the sum of the volumes of the sand and water measured separately? Explain.

OBJECTIVE ACCOMPLISHED? FIND OUT.

1. What number is used as a base in the metric system?
2. Describe each of these metric units and compare each with the English unit which is its closest equivalent: the meter, centimeter, kilometer and liter.
3. Complete each of the following relationships: 5cm = _____mm; 200mm = _____cm; 800cm = _____meters; 6.7 = _____meters; 3 liters = _____ml; 15ml = _____cm^3.
4. What is the volume of a box 3cm wide, 5cm long and 4cm high?
5. Suppose that you had a soft piece of clay shaped into a cube 2cm × 2cm × 2cm. How would you find its volume? If the piece of clay is rolled into a ball, how would you find its volume?
6. State whether it would be most convenient to measure each of the following in *meters, millimeters, centimeters or kilometers:* (a) the distance of a short running event in a track meet, (b) the distance between two cities, (c) the thickness of a hair, (d) your height.
7. Which would be more convenient to use, *liters* or *millimeters* to measure each of the following? (a) the amount of milk purchased at a grocery store, (b) medicine taken from an eyedropper, (c) the gasoline capacity of an automobile gas tank.

YOUR OBJECTIVE: To understand the difference between weight and mass; to find out how mass is measured; to understand what density means and how density is determined.

In the previous section you learned that one property of matter is that it takes up space. You also learned to measure this space. You will now learn about another property of matter — that it has weight. Matter also has mass. Although weight and mass are closely related, they are not the same thing. **Weight is a measure of the force of gravitational attraction between any two objects.** This force is called *gravity*. Most often weight refers to the force of attraction between the earth and some object on or close to its surface. But an object still has weight when it is on or close to the surface of the moon. Its weight there is due to the gravitational pull or attraction between it and the moon (FIG. 2-1).

The weight of an object changes as it is moved from one place to another. An object weighs less on the moon than it does on the earth. You will learn why this is so later on in this book when the law of universal gravitation is explained.

Mass is a measure of the amount of matter that an object contains. The amount of this mass depends upon the number and type of atoms that make up the object. If one object has twice as many atoms of one kind as another object, it then has twice as much mass. A piece of iron made up of 10 billion iron atoms has twice as great a mass as a piece of iron made up of 5 billion iron atoms.

The mass of an object does not change as it is moved from one place to another. An object has the same mass on the moon as it has on the earth. As long as atoms are not added to or taken away from the object, its mass remains constant no matter where it is taken in the universe.

The weight of an object, on the other hand, is determined not only by its mass but also by where it is located.

The standard unit for measuring mass in the metric system is the **gram (gm)**. The gram is defined as a mass equivalent to the mass of one cubic centimeter of water at a temperature of 4 degrees Celsius (4°C). You can get some idea of the mass of 1 gram if you remember that a nickel coin has a mass of about 5 grams. Very small amounts of mass are measured in milligrams (mg): 1 mg = 1/1,000 gram.

A gram is a small unit to use to measure the mass of relatively large objects such as people or automobiles. For such measurements the kilogram (kg) is more useful. The prefix *kilo* means 1,000 times. One kilogram, then, is equal to 1,000 grams. A kilogram mass weighs about 2.2 pounds at the earth's surface. (TABLE 2-1).

TABLE 2-1 Metric Mass Relationships	
1,000 mg	= 1 gram
1,000 grams	= 1 kilogram

FIGURE 2-1 Astronaut Edwin E. Aldrin, Jr. steps down to the moon's surface. The moon's gravity is less than the earth's. Will the astronaut's weight be the same, more, or less than his weight on earth? *(NASA)*

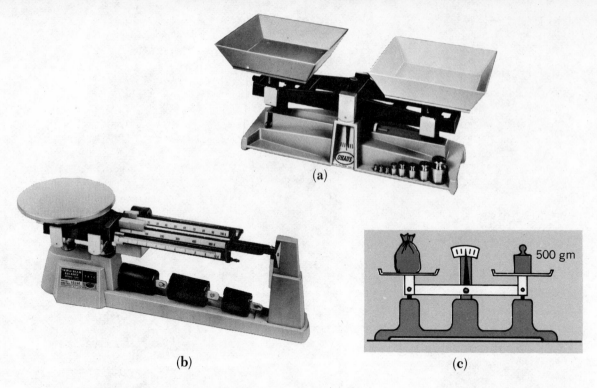

(a)

(b)

(c)

500 gm

FIGURE 2-2 (a) An even-arm balance (b) A triple beam balance: To find the mass of an object placed on the balance pan adust the sliding bars (riders) of the beams until the balance pointer (right) is level with the line indicating zero. First adjust the center beam (0 to 500 gm), then the rear beam (0 to 100 gm), and lastly the forward beam (0 to 10 gm). If an object weighed 455 gm, the center beam rider would rest at 400 gm, the rear at 50 gm, and the forward at 5 gm when the object is balanced. Additional weights (below the beams) may be attached for heavier weighings. (*Ohaus Scale Corporation*)

A *balance* is a device used to measure small masses by comparing them to a known mass. One kind of balance consists of a bar supported in the center that has two pans of equal mass attached at each end. There are other kinds of balances. Figure 2-2a,b shows two kinds often found in a school laboratory.

To find the mass of an object using a balance like the one in Figure 2-2a, you first place the object on the *left* pan and add weights (the known masses) to the right pan until the pointer indicates that

the pans are balanced (the pointer comes to rest at the center point of the scale as shown in Figure 2-2c). The total of the weights placed on the right pan gives you the mass of the object on the left pan. In English-speaking countries the *pound* is often used to measure mass. In the United States one pound mass is about 0.454 kilogram.

Has anyone ever asked you the trick question, "Which is heavier, a pound of lead or a pound of cork?" If you did not answer too quickly, and gave this ques-

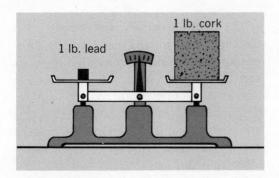

FIGURE 2-3

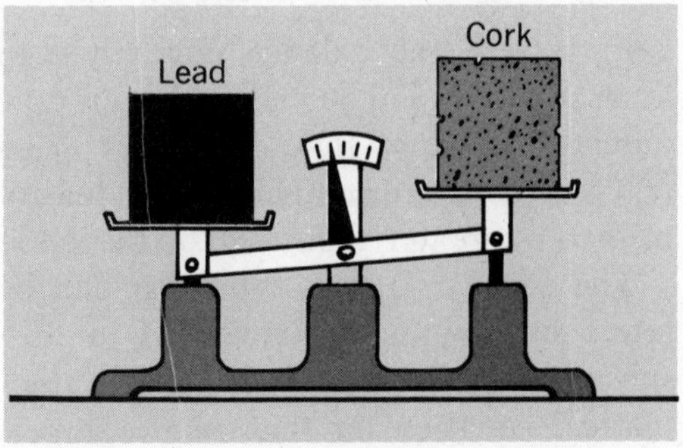

FIGURE 2-4

tion some thought, you probably answered correctly that they both have the same weight — one pound. Since they both have the same weight they should also have the same mass. If placed on opposite pans of a balance, the pound of lead and the pound of cork would balance (FIG. 2-3). What is it that is different between these two blocks of material? The answer, of course, is that one block is larger or takes up more space than the other. The point is that it takes a much larger *volume* of cork to balance the same mass of lead. Cork is less dense than lead. Another way of saying this is that lead has a greater *density* than cork.

Density is defined as the mass of a material for a given unit of volume. Suppose that *equal volumes* of lead and cork are placed on opposite pans of a balance (FIG.

BE CURIOUS 2-1:

Find out if you know how to use a balance to determine the mass of objects.

6 small solid objects
Red liquid
Blue liquid
4 containers
Balance

Before starting this investigation make sure that the balance is set so that the pointer is at zero on the scale, and that there are no weights on either pan. (If the balance you are using differs greatly from those described in text, obtain instruction in its use.) Never place liquids or powdery substances directly on the pans. Make all measurements to the smallest unit shown on the balance you are using.

a. Find the mass of each of the solid objects given to you for measurement. Record each mass.
b. *Estimate* the mass of some small objects such as coins, keys, and combs. Check your estimate by measuring their mass on the balance. Record each mass.
c. Find the mass of the liquids given to you by your instructor. Be sure that the mass of each container is not included in your results. See how many different methods you can use to find the mass of each liquid. Discuss these methods in class.

2-4). Why is the balance now tilted in favor of the lead? The mass of lead is now much greater than that of the cork when equal volumes are compared. The density of lead is greater than that of the cork.

The density (D) of a substance can be found by dividing the mass (M) of this substance by its volume (V). The mathematical equation for this relationship is written

$$D = \frac{M}{V}$$

FIGURE 2-6

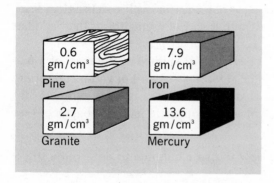

Pine — 0.6 gm/cm³
Iron — 7.9 gm/cm³
Granite — 2.7 gm/cm³
Mercury — 13.6 gm/cm³

The units for density will depend upon the units used for mass and volume. For example, if the mass is measured in *grams* and the volume in *milliliters*, then the density has units of *grams per milliliter* (gm/ml). If the mass is measured in grams and the volume in cubic centimeters, the density has units of *grams per cubic centimeter* (gm/cm³).

The density of different substances found on the earth's surface varies greatly (FIG. 2-6). The density of fresh water is one gram per cubic centimeter. Salt water from the sea has a slightly greater density. In the metric system the gram is defined as the mass of *one* cubic centimeter of water at 4°C. This definition automatically determines the density of water as one gram per cubic centimeter. Using the equation for density

$$D = \frac{M}{V}$$

$$= \frac{1.0 \text{ gm of water}}{1.0 \text{ cm}^3 \text{ of water}} = 1.0 \text{ gm/cm}^3$$

SAMPLE PROBLEM: Find the density of the object shown in Figure 2-5.

SOLUTION

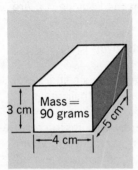

FIGURE 2-5

The volume is found by multiplying length, width and height.

$V = l \times w \times h$

$= 5 \text{ cm} \times 4 \text{ cm} \times 3 \text{ cm}$

$= 60 \text{ cm}^3$

The density is found by using the equation

$D = \frac{M}{V}$

$= \frac{90 \text{ gm}}{60 \text{ cm}^3}$

$= 1.5 \text{ gm/cm}^3$

BE CURIOUS 2-2: **Find the density of some solids and liquids.**

Balance
Wooden block
Metal block
Metric rule
Water
Graduated cylinder
Chunk of clay

Calculate the density of several objects by first measuring the mass and volume of each. Measurements of mass should be in *grams*. The volume of all solids should be in *cubic centimeters*, and the liquids in *millimeters*. Make all measurements to the nearest tenth of a unit whenever possible.

a. Find the *volume* of the wooden block and the metal block from their linear measurements. Determine the *mass* of each block. With this information find the *density* of each block.

b. Measure a *volume* of 100 milliliters of water in a graduated cylinder. How would you find the *mass* of this amount of water? Now calculate the *density* of the water. Repeat this process using 200 milliliters of water. How do the densities of the two different volumes of water compare? Explain your results.

✿ c. Use any convenient method to determine the density of a chunk of clay. Repeat, using about one *half* of the chunk of clay. How do the two results compare? Explain your results.

OBJECTIVE 2
ACCOMPLISHED?
FIND OUT.

1. How would you define each of these terms: *weight, mass,* and *density?*
2. Complete the following equations: 5,000mg = _____gm; 30 kg = _____gm; 8.2kg = _____gm; 48gm/ml = _____gm/cm³.
3. Which unit *grams, milligrams* or *kilograms* would be most convenient to use to measure the mass of each of the following? (a) apples (b) a pencil (c) a hair?
4. An object has a mass of 36 grams. It is 6cm long, 4cm wide, and 3cm high. What is its density?
5. If you took a solid rubber ball and squeezed it tightly, how would this affect its (a) *mass* (b) *volume* (c) *density?*
6. A football player has a mass of about 100 kilograms. How many pounds does he weigh? Would his weight be the same on the surface of the moon?

YOUR OBJECTIVE: To identify the three states of matter; and to describe how atoms behave in each of these states.

Matter is the stuff of which all objects in the universe are made. The scientist identifies **three different states in which matter can exist — solid, liquid, or gas.** In each of these states the particles (atoms and molecules) that make up a particular substance behave differently. A model of the structure of matter, which scientists have formed, will help you understand this behavior. The model pictures **all matter as being made up of very small energetic particles that are in continual**

State As used, refers to condition.
Particle A minute (tiny) part.

motion. And, that this motion can increase or decrease as *energy* is added or taken away from the substance.

Matter in the *solid state* has a definite shape, and retains this shape under normal conditions. A small iron block keeps its shape and form unless you heat it to extremely high temperatures that would cause it to melt. Under normal conditions the iron atoms have a limited amount of motion in the solid state and stay relatively close to each other. In all solid matter, the individual particles have much less energy of motion than in the liquid or gaseous states. You might compare the motion of particles in the solid state to students seated at their desks in a classroom. Although each student moves about in his or her seat, there is no wandering to other parts of the room. The motion is confined to a small region of each desk.

FIGURE 3-1 (a) A sorter separates diamonds. The darker stones will be used for tools. The white stones will become gems. (b) The crystalline structure of a diamond. *(Diamond Information Center, LaPine Scientific Company)*

Atoms in a solid may be arranged in an orderly fashion with respect to each other. When this is the case, the solid is described as having a *crystalline* structure. A crystalline substance that has a regular geometric shape is called a *crystal* (FIG. 3-1a,b).

Solid substances that do not have an orderly arrangement of their atoms are **noncrystalline.** Materials such as plastics or glass are good examples of noncrystalline substances.

A liquid does not have a definite shape. It takes the shape of the container in which it is poured (FIG. 3-2a,b,c).

The particles of a liquid substance move more freely than the particles in a solid. The particles in a liquid do not remain in the same place, but can move about to all parts of the liquid mass.

Since particles of a liquid are not held in place, you can stir and pour liquids. You can also mix two liquids together. **Substances in the liquid state can flow from one place to another.** For this reason liquids are called *fluids*.

When a liquid is poured from one container to another, its shape may change, but its volume remains the same.

Gases are also called fluids since they too flow from one place to another. Like a liquid, a gas takes the shape of its container. A gas, however, occupies *all* the space of its container. **The volume and**

(a)

(b)

(c)

FIGURE 3-2 (a) Observe that the shape of the liquid is determined by the glass. (b) The shape of the liquid changes as it is poured. (c) The shape of the liquid is determined by the goblet. *(Walter Schwarz)*

shape of a gas changes as the volume and shape of its container changes. The particles of gas are far more distant from each other than are particles in liquids or solids. Gas particles move about much more rapidly than do particles in a liquid or a solid.

The distance between particles in a gas is determined by the size of its container. If the size of the container is increased, the volume that the gas occupies will increase. The distance between gas particles will also increase. If the container is made smaller, the volume of the gas will decrease and so will the distance between gas particles.

OBJECTIVE 3 ACCOMPLISHED? FIND OUT.

1. What is matter? How do scientists describe the structure of matter?
2. What is the difference in the arrangement of atoms within a *crystalline* and a *noncrystalline* solid? What is a crystal?
3. What does *fluid* mean? Which states of matter are considered to be fluids?
4. Compare the abilities of solids, liquids, and gases to keep their volume and shape.
5. Compare the motion of particles within solids, liquids, and gases.

4 | MATTER CHANGES STATE

YOUR OBJECTIVE: To find out how and why matter changes state; and how mass is conserved when a change in state occurs.

A change in the temperature of a substance can cause that substance to change from one state to another. At low temperatures a substance may be in the *solid* state. As the temperature of this substance is raised, it reaches a point where it changes to the *liquid* state. If the temperature of the substance is raised still more, it will reach a point where the liquid will change to a *gas*. This process may be reversed by lowering the temperature until the substance returns to the liquid and then to the solid state.

The *Kinetic Theory* explains what scientists think happens when matter changes its state. **The kinetic theory states that as the temperature of a substance increases the particles (atoms and molecules) that make up the substance**

Kinetic Referring to motion.
Theory A reasonable, generally accepted explanation for something.

FIGURE 4-1 Particle motion: (a) solid (b) liquid and (c) gaseous state.

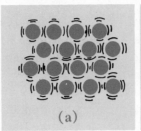 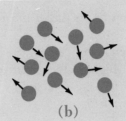 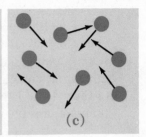

(a)   (b)   (c)

gain energy and move at greater speeds. And, that **the process is reversed when the temperature is decreased — the particles lose energy and move at lesser speeds.**

According to the kinetic theory, in the *solid* state the particles have low energy and their motion is limited. Particle motion is limited to vibration within a small range (FIG. 4-1a). You could say that particles in a solid vibrate "in place".

When the temperature of a solid reaches the melting point, the motion of the particles increases greatly. The particles have increased energy and can now move farther away from each other or *flow* (FIG. 4-1b). The substance is changed to the *liquid* state.

If you continue to add heat to a substance in the liquid state, a point will be reached when the particles will have sufficient energy and motion to change the substance into a gas (FIG. 4-1c). Gas particles are much farther apart and move at much greater speeds than do particles in the liquid state.

You can easily observe change of state in a common substance — water. Water

in the solid state is ice. If you take a beaker of ice cubes and heat it, the ice will melt to water in a short while. Water has **changed its state from solid to liquid by absorbing heat energy.** If you continue to add heat, the temperature of the water will rise until the boiling point is reached. The water then changes its state from liquid to gas. If you continue to heat the boiling water, it will soon disappear completely. It will change to an invisible gas that mixes with the air in the room.

When a substance changes from a solid to a liquid state, the process is called *melting*. Melting begins when enough heat has been added to a substance to raise its temperature to a certain point. This temperature is called the *melting point*. The melting points of a few substances are listed in TABLE 4-1.

TABLE 4-1	
Substance	*Melting Point °C*
Hydrogen	−259°C
Ethyl alcohol	−117°C
Tin	232°C
Oxygen	−218°C
Mercury	− 39°C
Aluminum	659°C

Vibrate **To move back and forth rapidly.**

The melting point of ice under normal conditions is 0°C (32°F). Some substances melt at much lower temperatures; others, at much higher temperatures (TABLE 4-1). The *minus sign* indicates temperatures below 0°C.

C stands for Celsius, a temperature scale on which the freezing point of water is marked as 0, the boiling point as 100. Divisions of the scale are decimals.

According to the kinetic theory of matter, **particles within a substance vibrate faster as the temperature of the substance increases.** When the temperature of a solid reaches a point at which the particles have enough energy to separate and move more freely, the substance becomes fluid rather than rigid. The substance is then in the liquid state.

When a substance changes from a liquid to a solid state the reverse happens. **As heat is removed from the liquid, the particles lose their energy (the temperature decreases), and their motion is**

BE CURIOUS 4-1:   **Find out what happens to the temperature of water as it changes state.**

Beaker
Ice
Water
Thermometer
Heat source
Watch or clock
Graph paper

Place a beaker of ice and water over a source of steady heat. Stir gently with a stirring rod as the ice is melts. Take temperature readings at one-minute intervals until the water begins to boil. Continue to boil the water and to take readings at one-minute intervals *for the next five minutes.* Also record the time when the last bit of ice disappears, and the time when the water begins to boil.

Make a graph from the data you have recorded. The temperature readings should be marked on the vertical axis of the graph and the time intervals on the horizontal axis. Also indicate on the graph the time at which the last bit of ice disappeared, and the time when the water began to boil. Join all the points on the graph with a continuous line. Study the resulting graph carefully and answer the following questions.

1. What happened to the temperature of the water during the time that the ice was melting?
2. What happened to the temperature of the water from the time when the last bit of ice disappeared to the time when the water began to boil?
3. What happened to the temperature of the water during the five-minute interval during which the water was boiling? What is your explanation of what you observed?
4. What is the melting point of ice as indicated by your graph?
5. What is the boiling point of water as indicated by your graph?

FIGURE 4-2 Note the whisps of vapor rising from the dry ice as some of the solid chunks change to a gas. Dry ice is solid carbon dioxide (CO_2). It has a temperature of about −79°C (−110°F). It is often called *snow ice*. Ice cream is often packed in dry ice for shipment. Dry ice is also widely used as a refrigerant. If you pick up a piece of dry ice it will burn your hand. *(Liquid Carbonic Corporation)*

slowed down. When the temperature has reached a particular point the particles no longer move freely. They vibrate in one place instead of moving freely from place to place. The change-over process — from liquid to solid — is called *freezing.* The temperature at which a substance changes from a liquid to a solid is called the freezing point. The freezing point and the melting point are the same. For example, water freezes or melts at 0°C.

When a substance changes from a liquid to a gas, the process is called *evaporation.* Evaporation can take place at various temperatures. But, the temperature point at which evaporation is rapid and the temperature remains constant (the same) is called the *boiling point.* The boiling point of water is 100° C. TABLE 4-2 shows the boiling point of some common substances. Note that the boiling point of oxygen and hydrogen are far below the freezing point of water.

When heat is removed from a gas the particles lose energy and slow down. The distance between particles decreases.

TABLE 4-2	
Substance	Boiling Point °C
Hydrogen	−253°C
Ethyl alcohol	78°C
Oxygen	−183°C
Mercury	357°C

When enough heat is taken away a change from the gas to liquid state takes place. This change of state is called *condensation.*

Some substances change *directly* from the solid state to the gaseous state without going through the liquid state. This is called *sublimation.* Perhaps you have seen what happens to "dry ice" as it absorbs heat. It sublimates. The solid dry ice changes directly into a gas (FIG. 4-2).

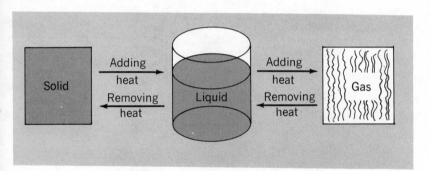

FIGURE 4-3 Heat must be added or removed from matter to change its state.

Perhaps you have also seen frost form on the outside of a window in very cold weather. If so, what you saw was water vapor (a gas in air) changing directly to a solid, crystalline state. These crystals disappear when they absorb sufficient heat. They return to a vapor (gaseous) state without going through the liquid (water) state.

In the investigation *(Be Curious 4-1)* you applied a steady heat to the water and ice mixture: that is, you added the same amount of heat throughout the entire time you applied heat. The graph of your temperature readings showed that during some intervals there was no temperature change. During other intervals the temperature rose. During the entire time a steady heat was being applied. The heat that was being absorbed during the interval when no temperature change took place was the heat energy required to change the state. During the melting interval a certain amount of heat had to be added so that a complete change of state would take place. The same thing happens during the boiling process. **A definite amount of heat energy is needed to change the state of a substance from solid to liquid, and from liquid to gas.** It can also be shown that the **same definite amount of heat energy must be taken away from a substance in order to reverse the process,** that is, to change the state from *gas* to *liquid,* and from *liquid* to *solid* (FIG. 4-3).

BE CURIOUS 4-2:   **Find out if mass is affected by a change of state.**

Ice cubes
Beaker or jar
Plastic wrap
Balance
Tools of your choice

Place a few ice cubes in a beaker or jar. Cover the container with plastic. Use a balance to measure and record the mass of the ice cubes. When all of the ice cubes have melted, measure and record the mass of the water. Compare the mass measurements for the solid and liquid states of the same quantity of water. Why was it necessary to cover the beaker? From your observations what conclusions can you make? Does the mass of a substance change when it changes from a solid to liquid?

You can now explain why it is that your body feels cool when you come out of the swimming pool — even on a warm day. Water on your skin is absorbing heat from your body as it evaporates (changes state from liquid to gas). Because your body is losing heat energy, you feel cool.

When you place an object in a refrigerator, the reverse heat energy exchange takes place. For example, if you put water in an ice cube tray and place the tray in a freezer, the water will change its state and form ice cubes. In the process of cooling the water loses heat. When sufficient heat energy is lost, the water reaches the freezing point and begins to change from liquid to solid.

When a substance changes its state mass is conserved. This means that a substance does not gain or lose mass as it changes from one state to another. When 100 grams of ice melts, it forms 100 grams of liquid water. And when 100 grams of liquid water evaporates, it forms 100 grams of water vapor. Similarly if the process were reversed, mass would not be lost or gained.

Conserved   Saved.

OBJECTIVE 4 ACCOMPLISHED? FIND OUT.

1. What causes a substance to change from one state to another?
2. What happens to the particles of a substance when its temperature increases? decreases?
3. What is the scientific definition of each of the following terms: melting, freezing, evaporation, condensation, boiling, sublimation?
4. What is the melting point of ice on the Celsius temperature scale?
5. How does the freezing point of water compare with the melting point of ice?
6. What is the boiling point of water on the Celsius temperature scale?
7. Is heat added to or taken away from a substance when it undergoes the following changes of state: melting, freezing, evaporating, condensing?
8. What happens to the mass of a substance when it changes from one state to another?
✳ 9. How does the Kinetic theory explain why a substance may exist in a solid, a liquid or a gaseous state?
✳ 10. Can there be a change in volume when a change of state takes place? Explain.

5 | FORCES AFFECT MATTER

YOUR OBJECTIVE: To understand how forces of attraction, such as cohesion and adhesion, can affect the properties of matter, such as capillary action and surface tension; to observe that application of force can deform matter.

Certain forces exist within solids, liquids and gases that attract one particle to another. As you found out in section 4, indi-vidual atoms or molecules are farther apart in a gas than in a solid or a liquid. Because the particles in a gas are separated by comparatively large distances, forces of attraction between the particles are *weak*. In solids or liquids the particles are much closer to one another and the forces of attraction between the particles are much *stronger*.

When forces of attraction exist between atoms or molecules that are of the

FIGURE 5-1   Water forms beaded drops on a greasy surface. Cohesion holds the water bead together. Adhesion holds the water bead to the greasy surface. *(Grant Heilman)*

same type, the attraction is referred to as **cohesion.** Cohesion explains why liquids such as water and mercury form droplets (FIG. 5-1). Forces of cohesion also explain why two solid pieces of the same substance can hold tightly to each other. When two pieces of steel are polished so that the surfaces are *very, very smooth,* the surfaces will stick together when brought together. Can you guess why the surfaces must be perfectly smooth?

When forces of attraction exist between atoms or molecules that are unlike, the attraction is referred to as adhesion. Adhesion explains why glue, paste and other kinds of adhesive can hold surfaces together. Why must you press down on the point of a pencil to write on paper?

Study the diagrams shown in *Be Curious 5-1.* The adhesion between water and glass particles is greater than the cohesion between water molecules. Because the forces of attraction are greater between water and glass than between water molecules, water molecules are attracted to the glass and are pulled up the sides of the container ever so slightly. If you look at a thermometer, you will see that the adhesion between mercury and the glass particles is less than the cohesion between mercury atoms. This is why mercury pulls away from the sides of its container ever so slightly.

Because of adhesion many liquids rise to considerable heights within thin tubes. **This process is called capillary action.** Tubes in which capillary action can take place are called *capillary tubes.* Water rises to the surface of soil by capillary action. Water rises within thin spaces between soil particles. The spaces act as capillary tubes. Water travels from plant roots to the leaves through long hollow fibers in the plant tissue in a similar way.

Compare the action of water in a glass capillary tube to the action of mercury in a capillary tube (FIG. 5-2a, b). Because the forces of cohesion between atoms of

> **Force** In physics, a push or a pull applied to an object.

BE CURIOUS 5-1: **See if you can find evidence that forces of attraction exist between particles.**

Water glass
Paper towel
Water

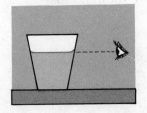

Slowly fill a clean, dry glass half full of water. Be careful not to splash water on the sides of the glass as you half fill the glass. Carefully place the glass on a desk or table top. Observe the sides of the glass where the water surface is in contact with the glass. Make sure your eye is even with the water level in the glass as in the figure to the left. Is the water level "flat" (level) where it meets the sides of the glass? Describe what you see where the water surface comes in contact with the glass. How can you explain what you see? What term describes what you observe?

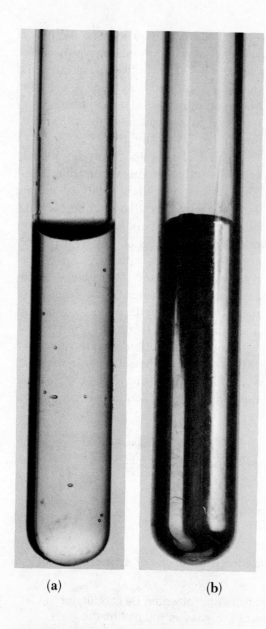

(a)   **(b)**

FIGURE 5-2 (a) Water rises where it meets the glass of the tube because in this case forces of adhesion are stronger than forces of cohesion between water molecules. (b) Mercury drops because forces of adhesion in this case are less than forces of cohesion between atoms. *(The Granger Collection)*

mercury are stronger than the forces of adhesion between mercury and glass, the mercury is forced *downward* in the tube.

Cohesion also causes a skinlike layer to form at the surface of a liquid. If a ring is dipped into water and withdrawn very slowly the water acts as if it had a thin skin along its surface. The ring stretches the skin until it is broken.

This force of attraction along the surface of a liquid is called surface tension. The amount of surface tension depends upon a number of things including the type of liquid used. For example, mercury has a greater surface tension than water. Soapy water has a lower surface tension than either mercury or pure water.

Surface tension explains why objects denser than water can rest on top of water when placed there gently. If the skinlike water surface is not broken, the denser object will rest on the surface in a shallow depression caused by its own mass. A razor blade is denser than water and therefore should sink when placed on the surface of water. If it is placed there gently, however, it will stay on the surface (Fig. 5-3). When the surface tension is broken by introducing a little detergent on the surface of the water, the blade will sink.

A bubble is a film of liquid in the form of a sphere. It is filled with air or some other gas. Particles that form the "skin" of the sphere are held together by surface tension. A bubble bursts when the force of cohesion among particles forming the

Depression   A low place.

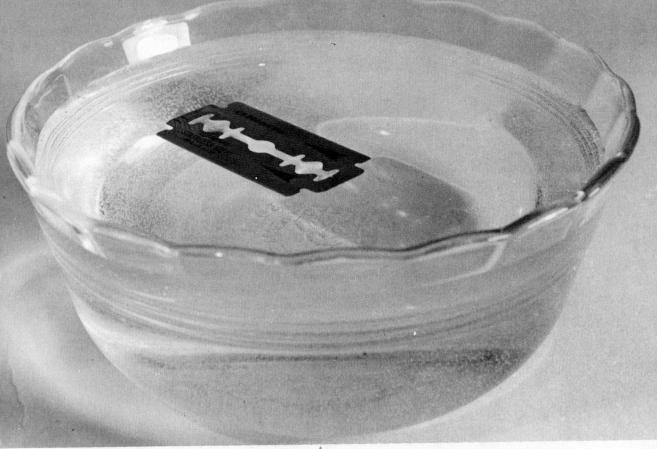

FIGURE 5-3 The cohesive forces of attraction between water molecules at the surface provide surface tension that supports the razor blade. *(Walter Schwarz)*

"skin" can no longer hold the particles together.

Matter is affected by forces exerted upon it. How much matter is affected by a force exerted upon it depends upon the amount of force exerted and the properties of the matter upon which the force is exerted. For example, compare the effect of pressure upon solids, liquids and gases. When atoms or molecules of a substance are forced closer to one another by some outside pressure, the substance is said to be compressed. Since the atoms of solids or liquids are already very close to one another, it is almost impossible to force them any closer by compression. Since molecules of a gas are comparatively far apart, they can be compressed (FIG. 5-4a, b). As the amount of pressure *increases*, the molecules of a gas move closer and closer together.

When atoms or molecules of a substance move apart and the volume of the substance increases, the substance is said to expand. Gases expand easily. Solids or

Exerted Put forth.

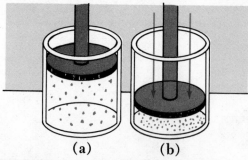

FIGURE 5-4 (a) A model of gas molecules within a cylinder. (b) As the piston is pushed down, the gas is compressed. What happens to the distance between molecules?

the same. **The density of a substance increases when it is compressed.** As substances expand their volume increases, but the mass also remains the same. **The density of a substance decreases when it is expanded.**

If you stretch a rubber band a small amount, it returns to its normal length and shape when you release it. When you bounce a basketball on a wooden floor, its size and shape changes as it strikes the floor. As it comes away from the floor it regains its original size and shape. **Objects that regain their original size and shape when a deforming force is removed are said to be elastic.**

A spring may be stretched so far that it cannot return to its original size and

liquids do not expand as easily. Solids and liquids normally expand only a small amount.

As substances are compressed their volume decreases, but their mass remains

BE CURIOUS 5-2:   **Find out how forces change the length of a spring.**

Spring
Set of small weights or
*   washers of equal mass*
Meter stick
Graph paper

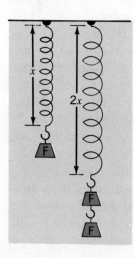

Suspend a light spring in some convenient place — such as a door handle. Measure and record the length of the spring. Hang a small weight or washer on the spring and again measure and record the length of the spring and the force applied to the spring. Continue to add weights or washers (of equal mass) recording the length of the spring and the amount of force used for each additional weight. (See figure to the left.) Repeat the procedure to obtain at least five readings of force.

Plot a graph of your data. Plot the data *Length of Spring* on the horizontal axis of your graph. Plot the *Force* used on the verticle axis. Study your graph carefully and answer each of the following questions.
1. What happens to the spring when force is applied?
2. What is the relationship between the force and the length of the spring?
3. Do you think that this relationship would continue if a very large number of weights were added?
4. How would the graph differ if you had used a stronger and stiffer spring in this investigation?

If time permits, repeat the investigation using a different spring.

✽ **SAMPLE PROBLEM:** A certain spring requires 6 newtons of force to stretch it 3 centimeters beyond its normal length. Find (a) the value of the constant k for this spring; (b) the force needed to stretch the spring 5 centimeters beyond its normal length.

SOLUTION:

a. The equation $F = kx$ may be written as

$$k = \frac{F}{x}$$

Then

$$k = \frac{6 \text{ newtons}}{3 \text{ centimeters}}$$

$$= 2 \text{ nt/cm}$$

b. The value of k obtained in part (a) and the value of x are substituted in the equation

$$F = kx$$

$$= 2 \text{ nt/cm} \times 5 \text{ cm}$$

$$= 10 \text{ newtons}$$

shape when the stretching force is removed. A plastic ruler can be bent so far that it cannot regain its original shape. It will break instead. When this happens the object has been stretched or bent beyond the *elastic limit*.

When you change the shape of an object, you *deform* the object. When you stretch rubber bands, bounce a basketball, stretch springs, bend plastic rulers and other elastic materials, you *deform* them. The amount of deformation depends upon the force that is applied to the object, and upon the kind of material being deformed. A law, known as *Hooke's law*, can be used to predict the amount of deformation resulting from a given force as long as the elastic limit is not reached.

In the first investigation of this section you found that a large mass caused the spring to stretch a greater distance than did a smaller mass. A straight line graph is a visual way of representing Hooke's law (Fig. 5-5). It shows you that if a force F stretches a spring a distance x, then a

FIGURE 5-5 The graph shows that the stretch of a spring increases as the force applied to the spring increases. Will the stretch continue to increase indefinitely?

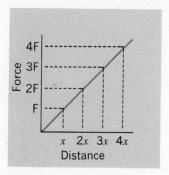

force twice as large (2F) is needed to stretch the spring twice the distance (2x).

 ☀ Hooke's law may be stated in the form of a mathematical equation $F = kx$. F is the deforming force, x is the distance of stretch in the case of a spring, and k is a constant that depends on the kind of material being deformed. Since the constant k from the equation is the result of divid-ing a force F by a distance x, it will have units like pounds per inch, or newtons per centimeter. A value of 5 newtons per centimeter says that a force of 5 newtons will stretch a spring 1 centimeter.

> **Constant**   Unchanging, the same.

OBJECTIVE 5 ACCOMPLISHED? FIND OUT.

1. Define cohesion and adhesion.
2. What is capillary action? Give one example of capillary action.
3. What is surface tension?
4. Explain why an object denser than water can "float" on water.
5. What happens to the volume of a gas when it is compressed? What happens to its mass? What happens to its density?
6. What kind of objects are said to be elastic?
7. What is meant by the term elastic limit?
8. How is amount of deformation related to force?
 ☀ 9. A certain spring stretched 4.0 cm when a force of 2.0 newtons is applied to it. Find the value of k for this spring. Plot a graph of *Force* versus *Distance Stretched* for this spring.

6 | MATTER EXERTS PRESSURE

YOUR OBJECTIVE: To understand different ways that matter can exert pressure; and to find out how that pressure can be calculated or measured.

Pressure is defined as force per unit area. Because solids have mass, solids exert pressure. **The amount of pressure exerted by a solid is determined by the volume and density of the solid.** A solid block resting on a table top exerts a force upon the table (FIG. 6-1). This force is spread over the area in which the block is in contact with the table.

Suppose that the block is turned so that a different surface touches the table (FIG. 6-1b). The block still exerts force — the *same amount* of force — but now, the force is spread over a *smaller area*. So, the amount of force exerted by the block per unit area — say on 1 square centimeter of table surface — is greater in this position

(FIG. 6-1b) than in the other position (FIG. 6-1a).

Similarly, if the area is kept constant and the force increases, the pressure also increases. The block in Figure 6-2a is exerting a pressure P on the surface below it because of its weight. If a similar block is added, the pressure will become twice as great ($2P$) (FIG. 6-2b). This is true because the weight has doubled, but the force is exerted upon the same area. As one block is piled upon another, the pressure increases (FIG. 6-2c).

Since fluids have mass, fluids also exert pressure. The amount of downward pressure exerted by a fluid at any point is determined by the height or depth of a column of this fluid, and by the density of the fluid.

As the depth in a fluid increases, the pressure also increases. As you dive deeper into a lake, the pressure on your body increases because you have a greater

FIGURE 6-1 (a) In this position the block rests on an area of 4 cm². (b) In this position the block rests on an area of 2 cm².

FIGURE 6-2 (a) If the pressure exerted by this block is represented by the letter P, how could you represent the pressure exerted by the block in (b)? in (c)?

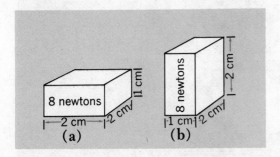

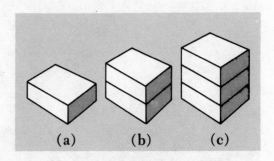

column of water above you (FIG. 6-3). If your body were beneath the same-sized column of mercury as of water, the pressure would be greater when you were beneath the mercury. Mercury has greater density than water.

As the density of a substance increases, the pressure that it exerts also increases.

A straight-line graph such as this tells you that the pressure will be *twice* as great if the density *doubles* (FIG. 6-4). The pressure will be *three* times as great if the density *triples*. Mercury is 13½ times denser than water. The pressure at a certain depth of mercury, then, is 13½ times greater than in the same depth of water.

FIGURE 6-3   Why must this deep-sea diver wear a pressurized suit? (*U.S. Navy*)

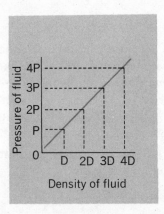

FIGURE 6-4

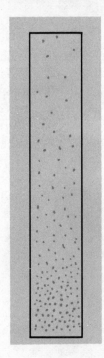

FIGURE 6-5 This model of a tall column of gas shows that the molecules at the bottom are pushed closer together by the weight of the molecules above them.

Note that the downward pressure of a fluid is determined only by the depth and density of the fluid. The pressure does *not* depend on the shape or volume of the fluid. The pressure one foot below the surface of the water in a bathtub is the same as the pressure one foot below the surface of a lake.

Since liquids compress very slightly even at extremely high pressures, the density of a liquid remains almost the same or constant at any depth. Gases, on the other hand, are easily compressed. As a result the density of a gas does not remain constant at various depths. The density of a gas increases as the depth of the gas increases (FIG. 6-5).

The *kinetic theory of matter* states that the molecules of a gas move at high speeds. When a gas is enclosed in a container the moving molecules collide with the walls of the container. These collisions cause a pressure on the container walls. If a gas is enclosed in a container, its pressure will tend to increase when the molecules hit the walls harder. This

happens when the molecules move with greater speed as the temperature of the gas is raised. The pressure of a contained gas increases when it is heated.

The pressure of a gas also increases when you reduce the volume of the gas. When you reduce the volume, there is less space for the molecules to occupy. The molecules have smaller distances to go to bump into one another and into the confining walls. This greatly increases the number of collisions among molecules and between molecules and the walls. The total effect is to increase the pressure of the gas on the walls of the container.

1. What is pressure?
2. What determines the amount of pressure exerted by a solid?
3. Suppose that you are swimming under water. What two factors determine how much downward pressure the fluid water exerts on your body?
4. What happens to the pressure of an enclosed gas when its *temperature increases?* When its *volume decreases?*
✴ 5. Explain why the density of a gas varies greatly as depth changes and why this does not happen to a fluid.

7 | FLUIDS AT REST

YOUR OBJECTIVE: To understand and apply Pascal's and Archimedes' principle.

Pascal's law is named after the French scientist, Blaise Pascal who lived in the seventeenth century. His investigations with pressure in fluids led him to conclusions usually referred to as Pascal's law.

Pascal's law states that **any pressure applied to a confined fluid is transmitted without loss of pressure to every part of the fluid.** The meaning of this statement becomes clear when you look at the hydraulic jack shown in Figure 7-1. A force applied *downward* on the right-hand pis-

FIGURE 7-1 A cross section of a hydraulic jack.

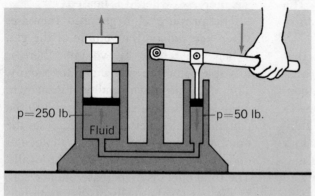

p=250 lb.

Fluid

p=50 lb.

ton results in an increase in pressure which is exerted in *all parts of the liquid equally. Each square inch* of surface of the walls confining the liquid receives the added pressure in the *same amount.* Note that the area of the left-hand piston is five times larger than the right-hand piston. If the pressure is transmitted without loss of pressure, then a 1 pound force applied to the right-hand piston will move the left-hand piston upward with a force that is five times greater. This explains why hydraulic devices such as hydraulic lifts at a service station can raise heavy objects with relatively small force (FIG. 7-2).

You know that some objects float on water while others sink. If you swim, you also know that objects appear lighter when underneath water. About the year 250 B.C. the famous Greek scientist and inventor Archimedes investigated these facts and explained why this is so.

Archimedes immersed a number of objects in water and observed what happened. He came to the conclusion that there is a force in water that pushes up-

FIGURE 7-2   The auto mechanic pushes down on a lever that transmits this force to a piston of a hydraulic jack as shown in Figure 7-1. Compressed air is the fluid in this hydraulic lift. The jack is beneath the garage floor. *(Ray W. Jones)*

ward on floating or submerged objects. This force that is present in all fluids is now called *buoyant force.* Archimedes said that in some cases the buoyant force is great enough to cause objects to float. In other cases, it is not great enough to cause objects to float, but will make the objects appear to weigh less. Archimedes' principle states that **this buoyant force on an object placed in a fluid is exactly equal to the weight of the amount of fluid the object displaces.** This is true whether the object floats or sinks in a fluid.

If you carefully lower a small wooden block in water until it floats and collect the water displaced by the part of the block that is under water you will discover that the water displaced *weighs the*

> **Submerge** To place under or into a fluid.

same as the wooden block. **The weight of the fluid displaced by a floating object is equal to the weight of the object.**

An object floats in a fluid if its density is the same or less than the density of the fluid. Wood usually floats on water because the density of most woods is less than the density of water. Iron can float in mercury because mercury is a liquid that is denser than iron. A helium-filled balloon floats in air because its density is less than the density of air.

If you lower an iron block that has a volume equal to 10cm^3 into water, its weight will force it to the bottom of the container. Since the entire volume of the iron block is under water, the block displaces 10cm^3 of water. According to Archimedes' principle, **the buoyant force pushing upward on the iron block should be equal to the weight of the water displaced.** In other words, the iron block

Legend has it that the development of Archimedes' principle began when the king of Syracuse (a city in ancient Greece) ordered a crown of pure gold made for himself. After the crown was made and delivered to him, the king wondered if the crown was really pure gold. The king summoned Archimedes and asked if he could determine whether the crown was indeed made of pure gold. Archimedes promised to do this without damaging the crown.

One day, while at a public bath, Archimedes observed that the water level rose when he got into the bathtub. This gave him an idea as to how to find out whether the king's crown was pure gold. Archimedes

was so excited with his sudden inspiration that he left his bath and ran naked through the streets shouting. "Eureka! Eureka!" (I have found it! I have found it!)

Archimedes idea was to place the crown in water and measure the volume of water that it displaced. Using this volume and a measurement of the weight of the crown, he would calculate the crown's density. If he found this density to be less than the density of pure gold, he would know the crown was not pure gold. When Archimedes made his measurements and calculations he found that the density of the crown was not that of pure gold. The king had indeed been cheated by the maker of the crown.

should appear to weigh *less* under water, and this *apparent loss* of weight should be equal to the weight of the water displaced. If you attach a small iron block to a spring scale and weigh the block then submerge the block in water and weigh the block again, you will find the block submerged in water weighs less. If you weigh a volume of water equal to 10cm³, you will find that the difference in the weight of the block in air and the weight in water is equal to the weight of the volume of water.

A 10cm³ block of iron has a mass of about 78 grams. When submerged in water the block displaces a volume of water equal to 10cm³. Since a 10cm³ volume of water has a mass of 10 grams, the iron block appears to be 10 grams less when it is in water. The mass of the iron block appears to have been reduced to about 68 grams (Fig. 7-3).

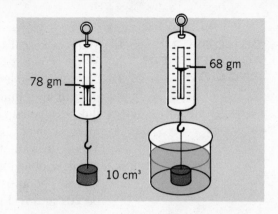

FIGURE 7-3 The apparent loss in mass is equal to the mass of the water displaced by the iron block.

For objects that do *not* float, Archimedes' principle is stated: **the volume of the fluid displaced by a submerged object is equal to the volume of the object. The apparent loss of weight by the submerged object is equal to the weight of the fluid displaced.**

BE CURIOUS 7-1:   **Find out if Archimedes' principle is true.**

Part A

Cork
Wooden block
Oveflow can
Catch bucket
Balance
Metal blocks
Spring balance

Use the cork and wooden block and the method described in this section to verify Archimedes' principle for *floating* objects. Record the mass of each object and that of the displaced water in grams. How does the mass of each object compare with the mass of the water that it displaced?

Part B

Use the metal blocks and the method described in this section to verify Archimedes' principle for objects that *do not* float. Record the weight of each object out of water and its apparent weight in water. How do the two weights compare?

✻ Can you find a way to determine if any apparent weight loss you observed equals the amount of water displaced? If so, try it. What are your conclusions?

1. What is Pascal's law?
2. What is meant by the term buoyant force?
3. Complete each of the following statements. (a) An object *floating* in a fluid displaces. . . . (b) An object that *sinks* in a fluid displaces. . . . (c) The apparent loss of weight by a submerged object is equal to. . . .
4. When does an object float in a fluid?
✴ 5. An iron object with a volume of 30 cm^3 is submerged in water. The apparent mass of the object in water is found to be 204 grams. Find the actual mass of the object.

8 | FLUIDS IN MOTION

YOUR OBJECTIVE: To find out what happens to the volume of materials when they are heated; to understand how fluids are put in motion when heated; and to understand viscosity and Bernoulli's principle.

Matter usually expands when heated and contracts when cooled. This means that mass undergoes a change in volume as its temperature increases or decreases. Expansion or contraction of matter caused by temperature changes does not change the total mass of a substance.

Engineers and architects allow for the expansion or contraction of matter when they design roads, bridges and buildings. Expansion joints are used to prevent cracking and buckling when temperatures change.

Expansion and contraction are most noticeable in gases. When a volume of air inside a balloon is heated, the volume of air grows larger — the air occupies more space — and the balloon expands. When the air cools the volume of air becomes smaller — occupies less space — and the balloon contracts (FIG. 8-1).

Solids also expand and contract when heated and cooled. But, the changes in volume are not nearly as great as corresponding changes in gases. *A ball and ring* device is often used to demonstrate expansion of solid objects (FIG. 8-2). At room temperature the metal ball can eas-

Device Something constructed for a specific purpose.
Contract To reduce in size.

FIGURE 8-1 (a) The air in the flask at room temperature. (b) The air in the flask expands into the balloon when heated. (c) The air and balloon contract when cooled.

FIGURE 8-2 A ball and ring used with a heat source to show how metal expands when heated. *(LaPine Scientific Company)*

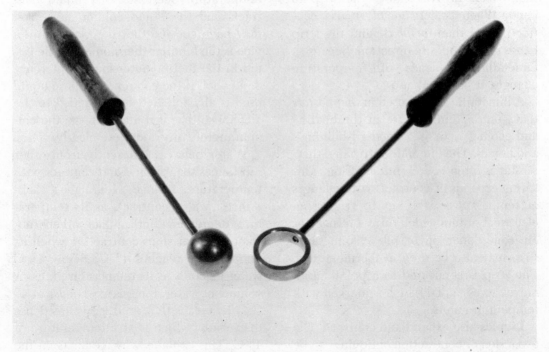

ily be passed through the metal ring. But, when the *ball* is heated in the flame of a burner, it expands and cannot pass through the ring. However, if in turn the *ring* is heated the ball can easily be passed through the ring.

All solids do not expand at the same rate when heated. A *bimetallic strip* is often used to illustrate this fact. A *bimetallic strip* is a metal bar made up of *two* different metals, which are tightly joined together (FIG. 8-3). When a bimetallic

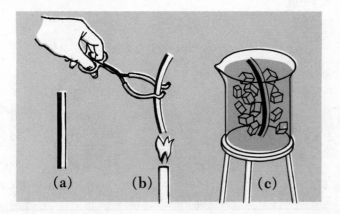

FIGURE 8-3 (a) A bimetallic strip at room temperature. (b) At a higher temperature. (c) At a cooler temperature.

strip is heated, metal A *expands* more than metal B. This causes the strip to bend. When cooled, *metal A* also *contracts* more than *metal B* and the strip bends in the opposite direction. Because a bimetallic strip bends with temperature changes, it has many uses.

A bimetallic strip is used in *thermostats* that control temperature in the heating and cooling systems of modern buildings and homes. The bimetallic strip in a thermostat is often coiled into a spring. This spring expands or contracts with changes of temperature. It is set to trigger an electrical system which turns a furnace or air conditioner on or off. A bimetallic strip may also be used in thermometers. The strip is connected to a pointer that moves across a dial which is marked with temperature units.

Liquids also expand and contract. The expansion or contraction of liquids can be observed with the use of a glass bulb-tube device. The glass bulb is filled with a liquid that expands as its temperature increases. The expansion forces the liquid up the tube. It is a simple matter to read from the scale the rise of the liquid in

degrees of temperature change. When temperature decreases, the liquid contracts and the liquid column falls. You may have guessed that the glassbulb device is the familiar thermometer. The liquid in the thermometer may be mercury or alcohol. If the column is red, the liquid used is alcohol. Red dye is added to the alcohol so that you may observe the column more easily and more clearly.

Water behaves differently from other substances that expand and contract with temperature change. Like most substances, water **contracts** as its temperature **decreases.** But, unlike other substances, **water stops contracting when its temperature reaches 4° C. Below 4° C water expands as its temperature falls.** A volume of water, then, is *most dense* at 4° C. This means that solid ice is less dense than water which is at a temperature of 4° C. This explains why ice floats on the top of a lake or pond in cold wintry weather. Can you think of some important consequences of this behavior?

Suppose that a glass container filled with water is heated. As the water absorbs heat from the heating element, that portion of the water nearest the heat will become hot first. Since this hotter water is less dense than the cooler portions at the top of the container, the hot water will be forced upward as the denser, cooler water sinks down. Until all of the water reaches the same temperature, circulation continues as *cooler* water *sinks* toward the *bottom,* and *warmer* water is pushed *upward.* **Circulation in a fluid that is caused by temperature differences is called convection.**

Convection also occurs in gases. When a heat source is placed under a glass chimney, the air near the heat source expands as its temperature rises (FIG. 8-4). This less dense, heated air is pushed up the chimney as the denser cool air moves in beneath.

Air circulates in the atmosphere because of convection. Warm air rises into

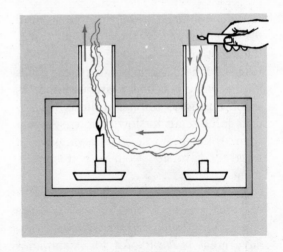

FIGURE 8-4 Apparatus like this is often used to demonstrate convection. When a candle is burning beneath one of the chimneys and a smoking object is held above the other chimney, smoke from the smoking object will be carried down into the box and then across the box and up through the chimney over the lighted candle. Explain why this happens.

Circulation **A moving around or through something returning to the starting point.**

the atmosphere as denser cool air falls to lower levels pushing upward the warmer, lighter air. **Water circulates in lakes and oceans primarily because of convection.**

BE CURIOUS 8-1:   **Observe the circulation of water by convection.**

Large pyrex container
Electric heater
Ice cubes, water
Plastic bag, ink

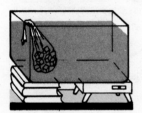

Place a source of heat under one end of a large water-filled container. (See figure on the left.) An electric hotplate turned to low heat is a good source to use. Place ice cubes in a plastic bag and suspend the bag at the other end of the container. Place a few drops of ink in the water and observe the motion of the colored water as heating takes place.

Make a drawing of the container showing the location of the heat source and of the ice. Use arrows to show the direction of the water circulation observed from the movement of the colored water. Explain why the colored water circulated as it did. What was the purpose of placing the bag of ice cubes in the water?

When objects move through a fluid they experience a force which resists their motion. The same resistance occurs when a fluid flows past an object. **This kind of force is called a viscous force.** As an object moves through a fluid, this viscous force tends to slow it down. As you swim through water in a pool, viscous force tends to slow your forward movement.

Some fluids, like thick syrup, tend to exert a larger viscous force than does water. A fluid such as thick syrup is said to have great *viscosity*. Air for example is viscous but not nearly as viscous as water. Water is viscous but not nearly as viscous as syrup.

Liquids have higher viscosity than gases. The *"thickness"* of a liquid determines its viscosity. The *temperature* of a fluid is a factor in determining the degree of viscosity of a fluid. **Viscosity increases as the temperature of the fluid decreases.** Motor oil, for example, becomes thicker and more viscous as it gets colder. If you chill maple syrup, it becomes thicker.

Viscous forces increase rapidly as the speed of an object moving in the fluid increases. You can experience viscous force when you try to walk in waist-high water. If you try running through waist-high water, you would find that the viscous force increases greatly.

Suppose that water is flowing out of a pipe at the rate of 10 gallons per minute (10 gal./min.). If the water continues to flow at this rate, you can assume that the water is flowing at this rate in *all sections* of this pipe. If the pipe has a narrow section, the water must be flowing at the rate of 10 gal./min. in this section also (FIG. 8-5).

If you think about this for a moment it will become clear that in order for the water to flow at the same rate of 10 gal./min. in the narrow section, it will have to flow faster at this point than in the wider sections of the pipe. It will have to flow faster if it is to move 10 gal./min. through the smaller pipe. In other words, the water must flow *faster* in the narrow sections and *slower* in the wider sections to move the water at the *rate* of 10 gal./min.

When a fluid (gas as well as liquid) moves through pipes or tubes, its speed increases in the narrow sections and de-

FIGURE 8-5 The same amount of water flows through all sections of the tube. Where does it flow faster?

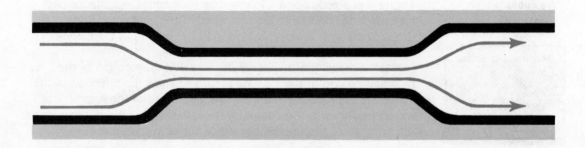

(a)

(b)

FIGURE 8-6   (a) Water flows from the nozzle of a garden hose. (b) A finger reduces the size of the opening and the water flows faster. If the water did not flow faster, what would happen to the hose? *(Walter Schwarz)*

creases in the wide sections. This is known as *Bernoulli's principle*.

Bernoulli's principle has many applications. The flow of a river is generally slow in its broad parts and fast in its narrow parts. You may have experienced Bernoulli's principle in the flow of water from the garden hose. The water flows out comparatively slowly from its open end when the nozzle is removed (FIG. 8-6a). If you reduce the size of the opening with your thumb, the water comes out with greater speed (FIG. 8-6b). Try to think of other examples of Bernoulli's principle.

BE CURIOUS 8-2:   **Find out how viscosity varies in different fluids.**

Part A

Water, alcohol, cooking oil
"Summer weight" motor oil
"All weather" motor oil
2 graduated cylinders
2 marbles
Piece of clay

In this part of the investigation you will compare the viscosity of different liquids. Pour one of the liquids on your list into a graduated cylinder until it is nearly full. Pour a second liquid from your list in the second graduated cylinder to the *same level*. Drop a marble in each of the liquids, *releasing them at the same time*. Note which marble reaches the bottom *first*. This is the liquid which has the *lower* viscosity of the two. Repeat this process with the other liquids on your list. Rank the liquids in order of their viscosity. Start with the least viscous liquid.

Part B

Using the method outlined in Part A, compare the viscosity of *warm* and *cold* "summer weight" motor oil. Repeat the procedure to compare the viscosity of *warm* and *cold* "all weather" weight motor oil. What are your conclusions?

Part C

Mold the chunk of clay into various shapes. Observe and compare the time that it takes each of the various shapes to fall through a certain height of water. What shape permitted the chunk of clay to fall fastest? Which shape slowed down its motion most. What is the effect of shape on the length of time taken for the clay to fall? Repeat the investigation using the two kinds of motor oil. Compare your findings.

OBJECTIVE 8
ACCOMPLISHED?
FIND OUT.

1. What happens to the volume of most substances when their temperature increases?
2. In which substances, solids, liquids or gases is expansion most noticeable? Why?
3. Explain why a bimetallic strip is used in thermostats and what its function is.
4. Explain the expansion and contraction of water as its temperature changes from 4°C to 0°C.
5. What causes convection to occur within a fluid? Describe convection in the atmosphere.
6. Does the motor oil in an automobile engine have a higher viscosity when it is *warm* or when it is *cold*? What is the reason?
7. State *Bernoulli's principle* in your own words.

IN THIS UNIT YOU FOUND OUT

Linear units in the metric system include the meter, centimeter, millimeter, and kilometer. Sometimes the volume of a solid object can be calculated from its linear measurements. The volume of ir-regular-shaped solids is often found by submerging them in water and noting the volume of water displaced.

The volume of a liquid can be found by pouring the liquid into a container that has graduations marked on its side. Met-ric units used to measure volume of liq-uids include the liter, and milliliter.

The weight and mass of matter are re-lated, but they are not the same thing. Weight is a measure between two bodies. Mass is a measure of the *amount* of matter contained by an object. The weight of an object depends on its mass and where the object is located. Mass is commonly mea-sured in metric units such as the gram, milligram, and kilogram. Weight is mea-sured in newtons in the metric system, and in ounces, pounds or tons in the En-glish system.

Density is the measure of the *mass* or weight of a substance per unit of volume. Density is calculated by dividing the mass or weight of the substance by its volume.

Matter can exist in three states — solid, liquid, and gas. Solids keep their volume and shape. Liquids keep their volume but take the shape of the container that holds them. Gases change volume and shape according to the shape and size of the container that holds them. Gases move more rapidly than liquids or solids.

The Kinetic Theory of Matter states that the atoms or molecules of a substance move faster as the temperature of the substance increases. In a solid, atoms have motion but remain in place. In fluids (liq-uids and gases) the atoms move freely from one place to another.

Heat must be added to a substance to cause it to melt or evaporate. Heat must be removed from a substance to cause it to condense or freeze. When a substance changes from one state to another, its mass remains the same.

The force between atoms or molecules of the same type is called cohesion. Co-hesion can cause surface tension along the surface of liquids. Adhesion is the force between atoms or molecules that are not alike. Adhesion can cause liquids to work their way up the inside of thin, hollow tubes or spaces. This action is called cap-illary action.

A substance is said to be compressed when its atoms or molecules are forced closer to each other. The density of a sub-stance increases when it is compressed. A substance is said to expand when its atoms or molecules are forced farther apart from one another. The density of a sub-stance decreases when it expands.

Substances whose size and shape are restored when a deforming force is re-moved are said to be elastic. Hooke's law can be used to predict how some elastic materials change size and shape when forces are applied.

Pressure can be calculated by dividing a force by the area on which the force acts. Pressure can be caused by the weight of a substance. The amount of pressure caused by the weight of a fluid increases with its depth and density. Pressure can also be caused by the motion of molecules in a gas as they strike the walls of its container. The pressure from moving molecules increases as the temperature increases and as the gas is compressed into a smaller volume. Similarly, the temperature increases as the pressure increases.

Pascal's law states that any pressure applied to an enclosed fluid is exerted in the same amount in every other part of the fluid. Archimedes' principle states that the buoyant force on an object immersed in a fluid is equal to the weight of the fluid displaced. This principle tells us that a floating object displaces its own *weight* of the fluid. An object that does not float displaces its own *volume* of the fluid, and is buoyed up by a force equal to the weight of the displaced fluid.

Most materials expand when heated and contract when cooled. Heating or cooling materials causes a change in their density. A fluid can be made to circulate because of density differences caused by heating or cooling. This circulation is called convection.

A viscous force is caused by a moving fluid. It is also caused by an object moving through a fluid. The thickness of a fluid determines its viscosity. The viscous force on an object moving through a fluid depends on the viscosity, and the speed and shape of the object.

Fluids flowing through a series of different size tubes flow faster in narrow tubes and slower through wider tubes.

UNIT OBJECTIVES ACCOMPLISHED? FIND OUT.

Part A   Match the numbered phrases in the left-hand column with the lettered terms on the right.

1. Uses the gram and meter as units of measurement.
2. The force of attraction due to gravity.
3. The amount of matter in an object.
4. A solid substance made up of atoms arranged in an orderly manner.
5. States that the molecules within a substance move faster as its temperature increases.
6. States that the distance a spring stretches depends on the force applied.

a. Archimedes' principle
b. Bernoulli's principle
c. crystalline
d. English system
e. Hooke's law
f. kinetic theory of matter
g. mass
h. metric system
i. Pascal's law
j. pressure
k. weight

7. The force per unit area.
8. Any pressure applied to an enclosed fluid will be felt in that same amount in every other part of the fluid.
9. The buoyant force on an object is equal to the weight of the fluid displaced by the object.
10. A liquid speeds up when it passes from a larger to a smaller tube.

Part B Choose your answer carefully.

1. Which of the following is *not* a unit of linear measurement (a) liter (b) meter (c) foot (d) kilometer?
2. If a classmate told you that his younger sister was one (a) meter (b) centimeter (c) millimeter (d) kilometer tall, it would seem reasonable.
3. Which of the following is *not* a unit of volume (a) quart (b) liter (c) cubic centimeter (d) meter?
4. If an object were taken from the earth to the moon its weight (a) would increase and its mass decrease (b) would remain the same and its mass increase (c) would decrease and its mass remain the same (d) would decrease and its mass decrease.
5. If a classmate told you that she has a text book that has a mass of one (a) gram (b) milligram (c) centigram (d) kilogram, it would seem reasonable.
6. The amount of space an object takes up is called its (a) mass (b) density (c) area (d) volume.
7. Which of the following is *not* true (a) solids tend to keep their own volume (b) solids tend to keep their own shape (c) solids have atoms that do not move *from place to place* (d) solids always have their atoms arranged in an orderly pattern?
8. Which of the following is *not* true (a) liquids tend to keep their own volume (b) liquids tend to keep their own shape (c) liquids have atoms that can move from place to place (d) liquids are called fluids?
9. Which of the following is *not* true (a) gases have molecules that move about rapidly (b) take the shape of their container (c) keep their own volume (d) are called fluids?

10. Heat must be added to a substance if it is to (a) melt or evaporate   (b) melt or condense   (c) freeze or evaporate (d) freeze or condense.

11. The freezing point of water is (a) 0   (b) 32   (c) 100   (d) 212 degrees Celsius.

12. The change in state from a solid directly to a gas is called (a) melting   (b) condensation   (c) evaporation   (d) sublimation.

13. The force of attraction between molecules of the same type is called (a) adhesion   (b) capillary action   (c) cohesion (d) viscosity.

14. If a substance is compressed its (a) volume decreases   (b) density decreases   (c) mass increases   (d) mass decreases.

15. Density is the measure of the mass or weight of a substance (a) at a certain pressure   (b) at a certain temperature   (c) at a certain viscosity   (d) per unit of volume.

16. The force of attraction along the surface of a liquid is called (a) adhesion (b) surface tension (c) capillary action (d) cohesion.

17. A certain object has a mass of 15 grams and a volume of 5 cm^3. If placed in water the object will (a) float   (b) displace 15 gm of water   (c) displace 5 cm^3 of water   (d) appear to weigh more.

18. The water pressure on the bottom of a lake is determined by the (a) shape and depth   (b) volume and shape   (c) depth and density   (d) volume and density of the lake water.

19. The viscous force on an object moving through a liquid does *not* depend on the (a) speed of the object   (b) viscosity of the liquid   (c) shape of the object   (d) depth of the liquid.

20. Convection currents are caused by (a) differences in density (b) the viscosity   (c) surface tension   (d) adhesion within fluids.

Part C   Think about and discuss these questions.

1. Suppose that a lake has a covering of ice. What is the temperature of the water beneath the ice? Give the reasoning for your answer.

✿ 2. Calculate the density of a 250 gm object if its volume is 10 cm^3.

✿ 3. A force of 10 newtons will stretch a certain spring a distance of 8 centimeters. How many centimeters will a force of 5 newtons stretch the spring?

✿ 4. Use an example to prove or disprove the statement: Mass is affected by a change of state.

5. Use an example to explain what capillary action is.

COMPOSITION OF MATTER

As scientists learn more and more about the particles that make up an atom, they use this understanding to produce new and potentially useful materials. The team of scientists shown in the unit opening photograph synthesized element 106. Element 106 is as yet undiscovered in the universe. The team is made up of physicists and nuclear chemists from the United States, Finland, and Germany. They conducted their research at the famous Lawrence Berkeley Laboratory at the University of California. In front is a model of the super Heavy Ion Linear Accelerator (HILAC) in which 106 was made.

1 | WHAT IS MATTER?

YOUR OBJECTIVE: To describe matter and the particles of which it is made; to understand how scientists use models to illustrate ideas.

The world around you is made up of many different things. Some of these things are called matter. **Matter can be defined as anything that has weight and takes up space.** Matter may be a *solid* such as rocks, leaves, or the steel of an automobile. It may be a *liquid* such as gasoline, mercury, or water. Matter may also be a *gas* such as the air you breathe. Solids, liquids and gases are called matter because they do have weight and take up space. Things such as sunlight, thoughts, and radio waves are not examples of matter. Why not?

Matter is either organic or inorganic. In its chemical structure, all organic matter contains carbon. Carbon is part of everything that is alive or that once has lived. For that reason *organic* matter is often referred to as *"living"* matter; and, inorganic matter is referred to as *"non-living"* matter. **Inorganic matter is matter that does not contain carbon.** Trees, horses and shoe leather are examples of organic matter. Glass, table salt, and iron are inorganic.

Chemistry is the science that deals with the make-up, or composition, of matter. It is concerned with any *changes* in the composition of matter and any *energy* that is related to such changes.

All matter is made up of very small "building blocks" called atoms. It would take over 100,000,000 atoms packed tightly together in a straight line to form a line about one centimeter long. Atoms are so small that they can be seen only with a very complex electron microscope. And when atoms are seen, it is not possible to see what they are made of (FIG. 1-1, page 52). If a scientist cannot see what atoms are made of, how is he able to know anything about them?

Suppose that there was an object in a closed cardboard box and you were not able to see it. Are there ways that you could know anything about the object without seeing it? Could you make some good guesses as to what it looks like without seeing it?

You could lift the box to see if the object in it were heavy or light. You could tilt the box to allow the object to roll or slide along the bottom. This would give you some idea of its shape. You could use a magnet to see if the object were made of certain types of metal. There are other tests that could help you find out many things about the unseen object. What tests would you suggest?

Since scientists cannot see an atom, they use similar kinds of tests to find out things about the atom. Some of these tests

Energy The ability to do work. Work is force exerted through a distance.

Centimeter Metric unit of length. Approximately equal to .4 of an inch.

FIGURE 1-1 This young scientist is examining a picture of nuclear particle reactions. The electronic measuring machine, which she is controlling, automatically plots the tracks of the particles that she selects and provides information for computer storage. In this unit you will find out about particles within an atom. *(Courtesy of Argonne National Laboratory.)*

include the shooting of "atomic bullets" through matter, passing atoms through strong electric and magnetic areas, and observing the action of one chemical on another. The information gained from tests, plus much mathematical figuring, has helped the scientist to make models of the atom.

A model is anything that represents something. A model can be a drawing, or a three-dimensional structure such as a model airplane. It can be a mathematical equation or a mental picture.

A model does not have to look like the thing that it represents. Usually models are very simple and show only the detail that is needed to make the model understandable. The "stick-figure" is a model of a person (FIG. 1-2). The stick-figure does not really look like a person, but it represents one. The "stick-figure" model helps to explain certain things about a person — that there are two arms and two legs attached to a body with a head on top. A "stick-figure" would probably not be a good model if you wanted to represent only a husky, brown-eyed athlete. Why not?

A *model* of the atom may be a picture of what the atom looks like. But, it is more important that the model explain certain observations that have been made about the atom. In some cases, the billiard ball model of the atom might be useful (FIG. 1-3a). If this model does not contain enough detail to explain certain things about the atom, the planetary model could be used (FIG. 1-3b). Why do you think that the names billiard ball and planetary were given to the two atomic models just mentioned?

The development of a model for the atom began several centuries B.C. when the philosophers of ancient Greece began to wonder what matter was made of. Some of them felt that all matter was made up of very small hard grains called atoms. There was no experimental evi-

Philosopher One who searches for understanding through logical reasoning rather than observation.
Billiard ball A small, hard ball.

dence at that time to suggest anything about the structure of the atom. The philosophers were very thoughtful men who used reasoning — with no experimentation — to come up with the idea that atoms were *indivisible.* By indivisible they meant that atoms could not be broken down into any smaller particle.

The Greek philosophers who thought that matter was made of atoms were called *atomists.* Democritus (di•ˈmäk•rət•əs) was one of the best known of the atomists. Democritus thought that all atoms were alike in that they were made of the same material. He also thought that

FIGURE 1-2

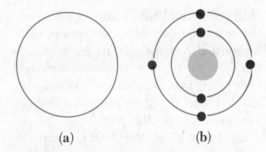

FIGURE 1-3 Two models of an atom (a) billiard ball (b) planetary.

atoms came in various sizes and shapes, which explained how they could make up all of the different kinds of matter found on the earth. It was a long time — more than 2000 years — before a new theory was suggested.

Theory A generally accepted explanation, based on evidence, of observable facts or phenomena.

A chemist takes matter apart and finds out what it is made of. A chemist could take a material such as water, and by using electrical energy find out that water is made up of two parts of hydrogen and one part of oxygen. She could also find out how much electrical energy was used in changing the water into hydrogen and oxygen.

Besides taking matter *apart,* the chemist puts it *together* to form new and useful material. The chemist is responsible for the development of many things in use today such as soap, plastics, synthetic fibers, and the heat-resistant materials used in space capsules.

Valeria Fisher, chief chemist of Bethlehem Mines Corporation, is responsible for

all analysis and research conducted in the company's laboratories.

✻ In 1803 John Dalton, an English scientist, proposed that different types of matter existed because the atoms of which the different types were made, were not alike. But, like Democritus, Dalton also assumed that the atom could *not* be broken down into smaller parts.

✻ By the end of the 1800's, evidence began to suggest that the atom *could* be broken down into smaller parts; that it was made up of small electrically charged particles. About 1900 J. J. Thomson developed a "raisin pudding" model of the atom (FIG. 1-4). His model consisted of an electrically charged positive (+) sphere with tiny negatively (—) charged particles embedded in it. The positive sphere represented the "pudding" of his model, and the negative particles the "raisins."

✻ At about the same time as Thomson was developing his "raisin pudding" model a French scientist, Henri Becquerel (bek•'rel), made a discovery which in turn was to supply the tool to probe more deeply into the structure of the atom. In 1896 while conducting experiments with a substance containing the element uranium, Becquerel discovered natural radioactivity. *Natural radioactivity* is defined as the ability of a nucleus to give off energy without external stimulation. Many scientists, including Polish-born Marie Curie and her French husband Pierre, investigated the nature of radioactivity. Among them, was a New Zealand-born scientist, Ernest Rutherford. Rutherford found that natural radioactivity was made up either of positively charged *alpha particles* or negatively charged *beta* particles usually accompanied by high-energy radiaton called *gamma rays.* Since alpha particles can

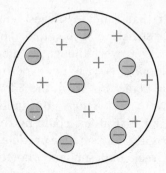

FIGURE 1-4   A "raisin pudding" model of an atom.

knock electrons loose from atoms through which they pass, the alpha ray — a ray that could be used as an "atomic bullet" — provided Rutherford with the tool that enabled him to find out that the atom was not a solid sphere, but mostly empty space. From his observations, Rutherford concluded that the atom consisted of a very small positively charged cluster of matter in the center of the atom with smaller negative particles traveling in orbits around the cluster.

✻ In 1913, a Danish scientist, Niels Bohr, further developed the idea that Rutherford had proposed. Bohr suggested that the negative particles could not be located at just *any* distance from the positively charged cluster, but, that negative particles orbit in distinct paths around the positive cluster (FIG. 1-5).

The modern atomic model shows a *swarm* or *cloud* of negative particles

Sphere   A globe or ball.
*High-energy radi-
ation*   Radiation of short
wave length, such as X rays.

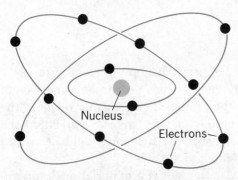

FIGURE 1-5 The Bohr atom shows the negative atomic particles in fixed orbits around the nucleus.

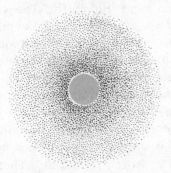

FIGURE 1-6 The present-day atomic model shows a cloud of negatively charged particles moving more freely about the nucleus than in the Bohr model. The orbits of the electrons are not fixed, but the region where electrons are most likely to orbit is defined by present-day theory. Note that the density of the cloud varies as the distance from the nucleus changes.

moving about a positively charged cluster (FIG. 1-6). The modernized model does *not* picture the negative particles in *orbit a fixed distance* from the positive cluster. Instead the negative particles are pictured as moving freely about the positive cluster and the model indicates the distance from the cluster where the negative particle *most likely* will be found.

Scientists who specialize in physics continue to investigate the structure of the atom. As they investigate, physicists find smaller and smaller particles within the atom. So far, the search for a simple, structureless, fundamental particle — the truly indivisible *atomos* — has failed. Each new level of investigation has shown evidence of structure and complexity within the particles observed. Some physicists have given up the search. They say there is no fundamental, indivisible particle of matter. Others say there is a fundamental particle and continue to search. Who is right? Only time will tell. But, for all "ordinary" purposes, atoms still remain the fundamental "building blocks" of matter.

> *Physics* The science that deals with the laws controlling motion, matter and energy.

OBJECTIVE 1
ACCOMPLISHED?
FIND OUT.

1. What is matter?
2. What is the difference between organic and inorganic matter?
3. If you take a course in chemistry, what will you study?
4. What is a model?
5. What are the building blocks of matter?
✴6. In what way do the atomic models of J. J. Thomson, Rutherford and Bohr differ from all of the earlier models?

YOUR OBJECTIVE: To find out what an element is; to identify and describe the three important particles of an atom; to find out in what ways the atoms of one element differ from the atoms of another element.

Scientists have identified about thirty different kinds of particles within an atom. In the study of basic chemistry, you need to know about only three kinds of particles — protons, electrons, and neutrons.

Protons have a positive (+) electrical charge. They are located in a cluster in the center of the atom. This cluster is called the *nucleus* of the atom. In most atomic models individual protons are not shown. Many times a number followed by the letter *P* represents the total number of protons that are present in the nucleus (FIG. 2-1).

The electrons have a negative (−) electrical charge, the opposite of that on the protons. Electrons are much smaller than protons. A proton is about 1,800 times heavier than an electron. If you use the proton as a standard and say that it weighs *one* unit, then you would say the electron weighs 1/1,800 of a unit (TABLE 2-1). Electrons orbit about the nucleus. Individual electrons are identified in some atomic models (FIG. 2-1).

Neutrons do not have an electrical charge. They are said to be *neutral*. Neutrons are found in the nucleus along with the protons. Like protons, individual neutrons are not shown in most atomic models. Instead, a number followed by the letter *N* represents the total number

of neutrons that are present (FIG. 2-1). The neutron has nearly the same weight as the proton. If you say the weight of the proton is *one* unit, then you would say the weight of the neutron would be *nearly one* unit (TABLE 2-1).

An element is a material made up entirely of one type of atom. All oxygen atoms are of the same type (FIG. 2-2a). All of the atoms of ordinary hydrogen are of the same type (FIG. 2-2b). The various types of atoms differ as to the number of electrons, protons, and neutrons found in the atom. Because of this, the element hydrogen is different from oxygen.

Protons and electrons are found in the atoms of all elements. Neutrons are found in the atoms of all but one element. The exception is the atom of the commonest form of hydrogen (FIG. 2-2b). There are three very important things to remember about the number of protons, electrons, and neutrons in an atom.

FIGURE 2-1 A lithium atom has 3 protons and 4 neutrons in the nucleus, and 3 electrons in orbit about the nucleus.

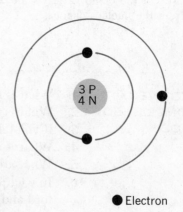

● Electron

FIGURE 2-2 (a) Oxygen atoms (b) hydrogen atoms. How do they differ?

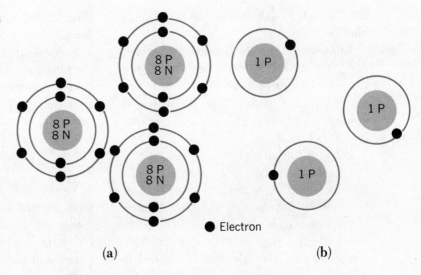

8 P
8 N

8 P
8 N

8 P
8 N

1 P

1 P

1 P

● Electron

(a)   (b)

TABLE 2-1			
Particle Name	Electrical Charge	Weight	Location in Atom
Proton	Positive (+)	1	Nucleus
Electron	Negative (−)	1/1,800	Orbit about nucleus
Neutron	Neutral (0)	1	Nucleus

1. All atoms of a certain element contain the same number of protons.
2. The number of electrons in an atom equals the number of protons in an atom.
3. The number of neutrons in an atom increases roughly as the number of protons and electrons increases.

Study TABLE 2-2. Compare the number of electrons and protons in each of the atoms of the common elements listed in the table. Note that the number of neutrons increases as the number of electrons and protons in an atom increases.

An element is a simple substance that cannot be broken down (separated) into

TABLE 2-2			
	Number of Particles in Each Atom		
Name of Element	Protons	Electrons	Neutrons
Hydrogen	1	1	0
Carbon	6	6	6
Aluminum	13	13	14
Iron	26	26	30
Uranium	92	92	146

any simpler substances by ordinary chemical means.

Salt water is *not* an element because it can be separated into simpler substances — water and salt. You could very easily separate the salt from the water. If you let the water evaporate, the salt will be left behind.

Water is *not* an element because it can be broken down into simpler substances — *hydrogen* and *oxygen.* This can be done by passing an electric current through a mixture of water and sulfuric acid and collecting the hydrogen and oxygen gases that are released. (The acid is added to the water because an electric current will not pass through pure water.) It can be shown that the hydrogen and oxygen did *not* come from the sulfuric acid since all of the acid remains after the water is gone.

The *salt* taken from the salt water is *not* an element. It can be broken down into *sodium* and *chlorine* by using ordinary chemical means. The hydrogen and oxygen (from the water), and the sodium and chlorine (from the salt) *are* elements. They *cannot* be broken down into any-

thing simpler by ordinary chemical means (FIG. 2-3).

Suppose that you were to take all matter found on the earth and break it down into its *simplest* forms. You would find that there are 92 *elements* which *occur naturally.* To date, *14 synthetic elements* have been made in the laboratory. Thus, 106 different elements are now known to exist.

You may wonder why the 106 elements have such different names. Usually elements are named by the people who discover them. Since the elements were discovered at very different times — in some instances hundreds of years apart — by various persons, the origin of the names varies greatly. For instance, the name of the element helium comes from the Greek word for sun, *helios.* The element was so named because it was first discovered in the atmosphere of the sun.

The names of synthetic elements sometimes honor a famous scientist. *Einsteinium* was named after Albert Einstein (FIG. 2-4a). The element *fermium* was named in honor of Enrico Fermi (FIG. 2-4b). Einstein and Fermi were outstanding physicists. Each received the Nobel prize for his special work in physics.

Synthetic elements have also been named after places where the experimental work leading to their discovery was carried out. *Europium* was named after the continent of Europe, and *berkelium* after the city of Berkeley, California. How do you think that the element *californium* got its name? Look the name up in an encyclopedia and find out.

Instead of writing the full name of each element chemists usually use an abbreviation — a *chemical symbol.* The chemical

FIGURE 2-3

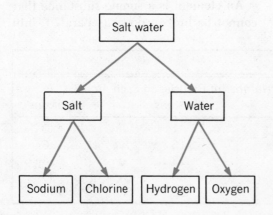

(a)

(b)

FIGURE 2-4 Two famous physicists: (a) Dr. Albert Einstein writing the equation he created to figure out the density (amount) of matter in the Milky Way, (b) Dr. Enrico Fermi, the leader of the first group of scientists to produce a scientist-initiated nuclear chain reaction *(Courtesy of Wide World Photos.)*

symbol for an element contains either one or two letters, but only the first letter is capitalized. **The chemical symbol stands for one atom of the element.**

In many cases it can easily be seen how the chemical symbol was obtained. It is the first or first two letters of the element name (TABLE 2-3).

Sometimes it will seem as if the symbol has no relationship to the name of the element. This may be because the symbol is an abbreviation of the foreign name of the element (TABLE 2-4).

Each element has its own atomic number. No other element can have that same atomic number. **The atomic**

TABLE 2-3

Element	Chemical Symbol
Hydrogen	H
Oxygen	O
Carbon	C
Cobalt	Co
Calcium	Ca
Cerium	Ce
Helium	He

TABLE 2-4

Element	Foreign Name	Chemical Symbol
Tungsten	Wolfram (German)	W
Sodium	Natrium (Latin)	Na
Mercury	Hydragyrum (Greek)	Hg

BE CURIOUS 2-1:
Given a few important numbers for several elements, can you arrange these elements in a table similar to Table 2-2?

From the data given at the end of this paragraph, figure out the number of protons each of the elements has. Decide which number of neutrons belongs with each element. Arrange the data in a table that is similar to TABLE 2-2. Place the element with the smallest number of protons first in the table. In what order will you place the remaining elements? Here is your data. 1. *Name of element plus electron number:* silver, 47, aluminum 13, lead 51, helium 2, neon 10. 2. *List of neutron numbers to match with elements* 13, 2, 10, 125, 60.

number is equal to the number of protons in each atom of that element. Oxygen has eight protons in each of its atoms, therefore, its atomic number is eight. No other element has or can have an atomic number of eight.

Since the number of electrons in an atom is the same as the number of protons, the atomic number is also the number of electrons in an atom. The element sodium has eleven protons and eleven electrons. So, its atomic number is eleven.

If you drove along the highway you would be surprised to see a road sign stating that the distance to the next city is 1,267,200 inches, or 3,168,000 centimeters. It is more convenient to say that the city is 20 miles away, or 32 kilometers away. When you state a measurement, you should always use the most convenient unit. In the examples given above the best unit would be either the mile or the kilometer (FIG. 2-5a,b).

> *Data* Facts or figures from which conclusions can be made.
> *Kilometer* Metric unit of length. Approximately .6 of a mile.

Because an atom has so little weight (mass), choosing the pound or the kilogram as a unit for measuring the weight (mass) of an atom would be even less convenient than measuring the distance between cities in inches or centimeters.

Atomic weight is a number given to an atom so that you can tell how heavy it is compared to other atoms. The unit that is commonly used when talking about the mass of an atom is the atomic mass unit, which is abbreviated a.m.u. If an atom has an atomic weight of 10 or a mass of 10 a.m.u., it means that the atom is twice as heavy as an atom with an atomic weight of 5 or a mass of 5 a.m.u. **The approximate atomic weight of an atom is equal to the total of the number of protons and neutrons in the nucleus.** The atom of aluminum has 13 protons and 14 neutrons so its atomic weight is *about* 27.

The atomic weights of most elements are usually *not given* as *whole* numbers. For simplicity, *you can round off atomic weights to the nearest whole number.* If you look at TABLE 2-5, you will see that the atomic weight of potassium is 39.102 a.m.u. You can "round" that number off to 39 a.m.u. To "round" off a number such as 39.102 you look at the numbers

(a)

(b)

FIGURE 2-5 Metric highway speed and distance markers have always been used in Europe. (a) This sign on the *Autobahn* (superhighway) near Norderstadt, Germany means: Save gasoline. Speed limit 80 km/hr. normal roads, and 100 km/hr. on *Autobahns*. (b) In another ten years all road signs in the United States will show metric units only. Some of the first metric/English road signs were installed on Interstate 71 in Ohio. *(Courtesy of United Press International.)*

TABLE 2-5							
Element	Symbol	Atomic Number	Atomic Weight	Element	Symbol	Atomic Number	Atomic Weight
Hydrogen	H	1	1.00797	Potassium	K	19	39.102
Helium	He	2	4.0026	Calcium	Ca	20	40.08
Boron	B	5	10.811	Iron	Fe	26	55.847
Carbon	C	6	12.01115	Copper	Cu	29	63.54
Nitrogen	N	7	14.0067	Zinc	Zn	30	65.37
Oxygen	O	8	15.9994	Silver	Ag	47	107.870
Fluorine	F	9	18.9984	Tin	Sn	50	118.69
Neon	Ne	10	20.183	Iodine	I	53	126.9044
Sodium	Na	11	22.9898	Platinum	Pt	78	195.09
Aluminum	Al	13	26.9815	Gold	Au	79	196.967
Silicon	Si	14	28.086	Mercury	Hg	80	200.59
Sulfur	S	16	32.064	Lead	Pb	82	207.19
Chlorine	Cl	17	35.453	Uranium	U	92	238.03

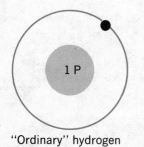  

FIGURE 2-6 Compare the number of neutrons in the nuclei of the three isotopes of hydrogen.

"Ordinary" hydrogen
Protium

Deuterium

Tritium

that are to the right of the decimal point. If the first number is 5 or greater than 5; you increase the number to the left of the decimal point by one number. Look at TABLE 2-5. The atomic weights of these elements would round off as follows: oxygen, 16; sulfur, 32; iron, 56; gold, 197.

✻ One of the reasons why the atomic weights for elements are usually not whole numbers is that atoms of most elements differ in the number of neutrons in the nucleus. So, the atomic weight of an element that has atoms with a varying number of neutrons shows the *average* of the atomic weights of these different atoms. Some hydrogen atoms contain *no* neutrons, some contain *one* neutron, and others contain *two* neutrons (FIG. 2-6). **Atoms of the same element that vary as to the number of neutrons in the nucleus are called isotopes of that atom.** Thus, hydrogen has three isotopes.

The standard for atomic weight is an isotope of carbon called carbon-12. The atom of carbon-12 has been assigned an atomic mass unit of 12 and all other atoms are compared to this standard. If an atom has an atomic weight of 24, it means that it weighs *twice* as much as an atom of carbon-12. An atom weighing *one-third* as much as an atom of carbon-12 would have an atomic weight of *four*.

OBJECTIVE 2
ACCOMPLISHED?
FIND OUT.

1. What is an element?
2. List the three important particles found within an atom. Describe each and tell where each is to be found within an atom.
3. A sodium atom contains 11 protons and 12 neutrons. How many electrons are in this atom?
4. What is meant by the term *atomic number?* How is it determined?
5. What is meant by the term *atomic weight?* How is it determined?

6. A certain atom has 15 protons, 15 electrons and 16 neutrons. What is its *atomic number?* What is its *atomic weight?*

✤ 7. What is an isotope?

3 | THE GROUPING OF ELEMENTS

YOUR OBJECTIVE: To recognize that there are groups of elements, which have similar characteristics; to understand how all elements are arranged in one table according to these characteristics.

Each element has its own unique atomic structure and therefore its own unique properties. You might wonder how chemists — whose work deals with the putting together and taking apart of elements to form new substances — know which element to use for what purpose. How do chemists keep track of the properties of 106 elements?

Although each element is unique, it may have properties quite similar to those of a number of other elements. Thus, it is possible to form groups of elements that have similar properties. **Scientists call groups of similar elements families.**

Elements have physical properties. **Physical properties are those that can be observed such as color and weight.** Elements also have chemical properties. **Chemical properties are those that have to do with the way one element joins another. Chemical properties of an element are determined by the atomic structure of the element.** The elements lithium, sodium, and potassium have similar properties. The identifying physical properties of the group are: they are shiny, fairly soft, light-weight metals (FIG. 3-1a, b, c);

Properties **Characteristics of an object or substance.**
Unique **Being the only one of its kind.**

Soft metal **One that can be cut with a knife.**

Metal **Elements that conduct heat and electricity; have luster (shine); can be shaped or molded.**

(a) (b) (c)

FIGURE 3-1 In what ways do these elements look like one another: (a) lithium, (b) sodium and (c) potassium? (*Courtesy of Fundamental Photographs from the Granger Collection.*)

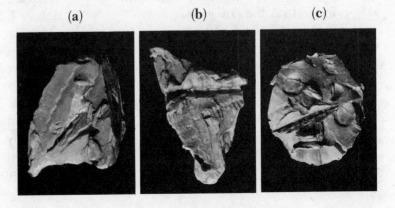

and, for metals, they melt easily (at low temperatures). Chemically, they react with the same substances to form new substances. (Because of atomic structure, every element cannot react with every other element.) For example, all three react with the element chlorine to form new substances called lithium chloride, sodium chloride, and potassium chloride.

Lithium, sodium, and potassium are members of a family called the alkali metals. The alkali metals react readily with many elements and substances. That is why in nature alkali metals are rarely found in the pure state. They are usually found joined with other elements. Other members of the group are rubidium, cesuim, and francium.

The elements helium, neon, argon, krypton, xenon, and radon have similar properties. They are gases that *normally* **do not react with any other elements.** So, at first the name given to this family was **the inert** (lazy) **gases.** Now that scientists have found that the inert gases can, under special conditions, react with some other elements, the inert gases are more often referred to as **the noble gases.** Because these gases are very scarce — they exist in only tiny, tiny amounts in the atmosphere — they are also commonly referred to as **the rare gases.**

Over a hundred years ago — when nothing was known about atomic structure or families — a Russian chemist named Dmitri Mendeleev (də•ˈmē•trē men•dəl•ˈā•əf) decided to organize the 63 known elements into a table. Atomic weights were known, so Mendeleev listed the elements in the order of increasing weights. Hydrogen had the lowest weight so he listed that element first.

As he listed the elements, Mendeleev noticed something interesting. Elements with similar properties appeared about every eighth time. When he observed this, Mendeleev arranged the table so that elements with similar properties could be grouped together. He did this by listing elements in horizontal rows. When he came to an element with properties similar to one already listed, he began a new horizontal row and placed the first element in this row in the same vertical column with the listed element(s) (above) that it resembled.

Mendeleev had a little difficulty with the elements tellurium and iodine. *The atomic weight of tellurium is greater than the atomic weight of iodine. Therefore tellurium should follow iodine in his table. But, when he placed iodine before tellurium, it was in a vertical row with dissimilar elements.* So, he reversed the position of iodine-tellurium to tellurium-iodine and all was fine. Each element then was like the other elements in its vertical row (FIG. 3-2). In his table, Mendeleev left spaces to indicate that elements would be discovered to fill those spaces. Placement of the space in the table predicted the kind of element that would eventually fill the space.

Although by modern standards Mendeleev's table is not completely correct, it is mostly correct. **The modern Periodic Table lists elements in order of increasing atomic number.** Since the list of elements in order of their atomic numbers is nearly

React In chemistry, to undergo change; to join with and/ or separate from other atoms.

Vertical column	Vertical column	Vertical column
74.9 As	79.0 Se	79.6 Br
121.8 Sb	127.6 Te	126.9 I

FIGURE 3-2 Selenium (Se) and tellurium (Te) are metal-like elements. Bromine (Br) and iodine (I) are nonmetals.

the same as that made in order of atomic weight, Mendeleev's table is much like the modern Periodic Table (pp. 18–19). Note the order of the elements tellurium and iodine in the modern table. Mendeleev's judgment — to reverse placement of these elements and disregard atomic weight — was correct.

If you study the Periodic Table, you will see that **members of the same family appear in the same vertical column.** All members of the family called *alkali metals* are found in the vertical column with *IA* above it. The family of *inert gases* is found in the column with *O* above it.

Why is the Periodic Table such a valuable tool? If you look at one small "square," you can see why (FIG. 3-3). **The chemical symbol (Al) in the center of the "square" tells you the name of the element (aluminum). The number in the upper left-hand corner is the atomic weight (26.9815) of the element. The number in the lower left-hand corner is the atomic number (13), which also tells you the number of protons in the nucleus of the atom.**

The smaller numbers in the upper right-hand corner tell you how many electrons are in each orbit of the aluminum atom. The number of electrons in orbit *nearest* the nucleus is listed first, then the next closest orbit and so on until the number in the outermost orbit is listed. Aluminum has two electrons in the shell nearest the nucleus, eight electrons in the next shell, and three in the outermost shell (FIG. 3-3). Compare this with the model of the aluminum atom (FIG. 3-4). What is the total

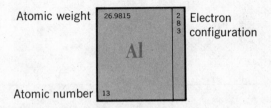

FIGURE 3-3 Look at the position of aluminum in the Periodic Table (pp. 18–19). Is aluminum a metal or a nonmetal? Will it have properties *strongly* resembling a metal or nonmetal?

FIGURE 3-4 Model of the electron configuration of the aluminum atom. A configuration is the relative arrangement of orbiting electrons to the central nucleus.

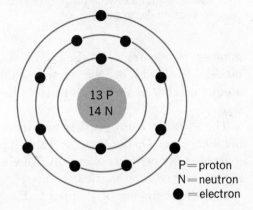

METALS

The Periodic Table of the Elements (legend): Atomic weight — Electron configuration — Symbol of Element — Atomic number, shown with example O: atomic weight 15.9994, atomic number 8, electron configuration 2 6.

Period	IA	IIA								
1	H 1.00797 (1) · 1									
2	Li 6.939 (2 1) · 3	Be 9.0122 (2 2) · 4								
3	Na 22.9898 (2 8 1) · 11	Mg 24.312 (2 8 2) · 12								
4	K 39.102 (2 8 8 1) · 19	Ca 40.08 (2 8 8 2) · 20	Sc 44.956 (2 8 9 2) · 21	Ti 47.90 (2 8 10 2) · 22	V 50.942 (2 8 11 2) · 23	Cr 51.996 (2 8 13 1) · 24	Mn 54.9380 (2 8 13 2) · 25	Fe 55.847 (2 8 14 2) · 26	Co 58.9332 (2 8 15 2) · 27
5	Rb 85.47 (2 8 18 8 1) · 37	Sr 87.62 (2 8 18 8 2) · 38	Y 88.905 (2 8 18 9 2) · 39	Zr 91.22 (2 8 18 10 2) · 40	Nb 92.906 (2 8 18 12 1) · 41	Mo 95.94 (2 8 18 13 1) · 42	Tc (99) (2 8 18 13 1) · 43	Ru 101.07 (2 8 18 15 1) · 44	Rh 102.905 (2 8 18 19 1) · 45
6	Cs 132.905 (2 8 18 18 8 1) · 55	Ba 137.34 (2 8 18 18 8 2) · 56	RARE EARTHS 57-71	Hf 178.49 (2 8 18 32 10 2) · 72	Ta 180.948 (2 8 18 32 11 2) · 73	W 183.85 (2 8 18 32 12 2) · 74	Re 186.2 (2 8 18 32 13 2) · 75	Os 190.2 (2 8 18 32 14 2) · 76	Ir 192.2 (2 8 18 32 15 2) · 77
7	Fr (223) (2 8 18 32 18 8 1) · 87	Ra 226 (2 8 18 32 18 8 2) · 88	ACTINIDE SERIES 89-103	NOT NAMED (257) · 104	NOT NAMED · 105	NOT NAMED (263) · 106			

The Periodic Table of the Elements The elements at the far left of the table are very reactive metals. As you move from element to element from left to right across the table gradually the properties of the elements change from metallic, to metal-like, to nonmetals, to gases.

number of electrons in the aluminum model? It should be the same as the atomic number of aluminum listed in the square.

The element hydrogen has an atomic number of *one*. On the Periodic Table it is located in the *vertical column* IA. All elements in that same vertical column are said to be members of *group* IA. Each hydrogen atom has one electron in orbit and one proton in the nucleus. Count the number of electrons in the model of the hydrogen atom (Fig. 2-7). Does the number agree with the atomic number in the "square" occupied by hydrogen on the Periodic Table?

The Periodic Table is made up of all known elements arranged in horizontal

NONMETALS

Periodic table section (each entry: atomic mass, symbol, atomic number, electron shells):

Symbol	Atomic Number	Atomic Mass	Electron Shells
He	2	4.0026	2
B	5	10.811	2, 3
C	6	12.01115	2, 4
N	7	14.0067	2, 5
O	8	15.9994	2, 6
F	9	18.9984	2, 7
Ne	10	20.183	2, 8
Al	13	26.9815	2, 8, 3
Si	14	28.086	2, 8, 4
P	15	30.9738	2, 8, 5
S	16	32.064	2, 8, 6
Cl	17	35.453	2, 8, 7
Ar	18	39.948	2, 8, 8
Ni	28	58.71	2, 8, 16, 2
Cu	29	63.54	2, 8, 18, 1
Zn	30	65.37	2, 8, 18, 2
Ga	31	69.72	2, 8, 18, 3
Ge	32	72.59	2, 8, 18, 4
As	33	74.9216	2, 8, 18, 5
Se	34	78.96	2, 8, 18, 6
Br	35	79.909	2, 8, 18, 7
Kr	36	83.80	2, 8, 18, 8
Pd	46	106.4	2, 8, 18, 18
Ag	47	107.870	2, 8, 18, 18, 1
Cd	48	112.40	2, 8, 18, 18, 2
In	49	114.82	2, 8, 18, 18, 3
Sn	50	118.69	2, 8, 18, 18, 4
Sb	51	121.75	2, 8, 18, 18, 5
Te	52	127.60	2, 8, 18, 18, 6
I	53	126.9044	2, 8, 18, 18, 7
Xe	54	131.30	2, 8, 18, 18, 8
Pt	78	195.09	2, 8, 18, 32, 17, 1
Au	79	196.967	2, 8, 18, 32, 18, 1
Hg	80	200.59	2, 8, 18, 32, 18, 2
Tl	81	204.37	2, 8, 18, 32, 18, 3
Pb	82	207.19	2, 8, 18, 32, 18, 4
Bi	83	208.980	2, 8, 18, 32, 18, 5
Po	84	(210)	2, 8, 18, 32, 18, 6
At	85	(210)	2, 8, 18, 32, 18, 7
Rn	86	(222)	2, 8, 18, 32, 18, 8

rows and vertical columns. All elements in the same horizontal row are said to belong to the same **period. Elements of the same period have the same number of electron "shells" or orbits surrounding the nucleus. The numbers in the right-hand column of each square show you how many "shells" each atom has and how many electrons are in each "shell".**

Elements found in the same vertical column are said to belong to the same **group. Elements of the same group have the same number of electrons in their outer "shell."** The *number* of electrons in the outer shell and the *distance* the outer shell is from the nucleus determine how that atom combines with other atoms.

Many scientists once thought that there would be no other elements than the 92 found in nature. Far-sighted scientists predicted that they would be able to *make* elements *beyond* uranium (atomic number 92), by adding protons to the nucleus of existing atoms. For example, if one more proton were added to the nucleus of a uranium atom a *new* element of atomic number 93 could be made. If two protons were added to the nucleus of

FIGURE 3-5 Aerial view of the main accelerator at the Atomic Energy Commission's (AEC's) National Accelerator Laboratory near Batavia, Illinois. The accelerator is 4 mi. (about 6½ km) in circumference and 1.24 mi. (2 km) in diameter. *(Courtesy of U.S. Atomic Energy Commission.)*

BE CURIOUS 3-1:   **What can you find out about the element potassium?**

Locate the element potassium (K) in the Periodic Table (pp. 18–19). Study the K "square" and its relationship to other elements in the Periodic Table. Find out:

a. What is the *atomic number* of potassium?

b. What is the value of its *atomic weight?* (Round it off to the nearest whole number.)

c. To what *period* does potassium belong? What do the atoms of the elements in this period have in common? List the symbols of other elements in this period.

d. To what *group* does potassium belong? What do the atoms of the elements in this group have in common? List the symbols of other elements in this group.

e. Draw a simple planetary model of a potassium atom. Show the number of protons and neutrons in the nucleus; and the number of electrons in each orbit of the atom.

uranium the result would be an element with an atomic number of 94.

Changing the makeup of the nucleus of an atom is a very complicated process. It normally involves bombarding the atom with atomic particles supercharged with energy. These particles can be protons, neutrons, or even the nucleus of one of the lighter atoms. Before 1930 there was no tool that scientists could use to give atomic particles enough energy to enter the nucleus of another atom. The situation was similar to a rifleman having a bullet (atomic particles of one atom) and a target (the nucleus of another atom), but no rifle to fire the bullet at the target.

In 1930 a "rifle" to fire the atomic "bullets" was built. The "rifle" was called a cyclotron (FIG. 3-5). The atomic particles placed in a cyclotron are made to go faster and faster until they have sufficient energy to enter an atomic nucleus. Then they are aimed at the target atoms. **When atomic "bullets" enter a nucleus with the proper amount of energy, they can add protons to that atom and increase the atomic number of the atom. Thus, a new element can be made** (FIG. 3-6).

FIGURE 3-6 This team working at AEC's Lawrence Berkeley Radiation Laboratory of the University of California synthesized the elements 104 and 105. The team, left to right: Matti Nurmia, a physicist from the University of Helsinki, Finland; James A. Harris, an American nuclear chemist; Kari and Firkko Eskola, physicists from the University of Helsinki; and Albert Ghiorso, the leader of the group. (*Courtesy of Lawrence Radiation Laboratory, Berkeley, California.*)

The first laboratory-made element was neptunium. Neptunium (atomic number 93) was produced synthetically in 1940 by Edwin M. McMillan and Philip Abelson at Berkeley, California. They produced the new element by bombarding uranium atoms with high-energy neutrons from a cyclotron.

The most recent elements produced are the elements 104, 105, and 106. Element 104 was first made in 1964 by a group of Russian scientists. The element was produced by bombarding plutonium with the nucleus of neon atoms. Further production of element 104 was reported by scientists at the University of California at Berkeley. In more recent years, the scientists in the Soviet Union and Berkeley, California have produced and examined the properties of elements 105 and 106.

Dr. Harris holds the target, californium-249, which was bombarded to create atoms 104 and 105.

Scientists predict that the Periodic Table will someday be extended far beyond element 106. Little is known about most synthetic elements. It is difficult to study some of them because so few atoms of the new element have been produced. Some of the synthetic elements exist only for a very short period of time after they are made, which makes it difficult to study these elements carefully.

Although the elements with atomic numbers above 92 are not considered to be *naturally-occurring* elements, many scientists think that they did exist on the earth at one time. Some think that these elements still do exist in the universe, but only in extremely small amounts — which makes them almost impossible to detect. Perhaps you will become a scientist and be one of a group who will find out if synthetic elements did or do occur naturally. It would be an exciting discovery!

OBJECTIVE 3 ACCOMPLISHED? FIND OUT.

1. What is meant by a family of elements? How are they located on the Periodic Table?
2. What is meant by the term *chemical properties* of an element?
3. What is the main difference between the way that Mendeleev organized his Periodic Table and the way the modern Periodic Table is organized?
4. What do all atoms in the same Group (vertical column) of the Periodic Table have in common?
5. What do all atoms in the same Period (horizontal row) of the Periodic Table have in common?

6. Use the Periodic Table (pp. 18–19) to find and list the atomic numbers and weights of oxygen (O), calcium (Ca) and zirconium (Zr).

☀ 7. Describe how elements with atomic numbers greater than uranium (92) can be made.

☀ 8. Why might you say that the Periodic Table is the "road map" of the chemist?

4 | CHEMICAL BONDS

YOUR OBJECTIVE: To understand how matter is made up of combinations of atoms held together by chemical bonds; to know how the atomic structures of elements determine which elements combine with one another.

All of the matter around you is made up of one or more elements. Some elements are found in their *pure* states. Gold and helium are elements that are found naturally in their pure states. **But, most matter is made up of combinations of elements.** Sugar is a combination of carbon, hydrogen, and oxygen. Table salt is a combination of sodium and chlorine. Hydrogen and oxygen combine to form water. Because elements can combine in many different ways, there are many different types of matter. How many different substances would occur naturally on the earth if the elements did not combine?

Atoms of different elements, or even of the same element, are held together by attractive forces called *chemical bonds*. **When a chemical bond exists between or among atoms the set of atoms acts (functions) as a unit. The bonding usually takes place in one of two ways: by transferring or sharing electrons. Both ways**

involve the electrons in the *outer* shell of an atom. When outer-shell electrons are **transferred** — given up by an atom and taken into the outer shell of another atom — **the type of bonding is called ionic. If outer-shell electrons** are not transferred from one atom to another but are **shared by two atoms (or groups of atoms), the bonding is called covalent.**

Atoms cannot bond in just any combination. **Atoms tend to combine in such a way that in the end their outer shells have eight or two electrons.** When an atom has **eight atoms** in its outer-shell, it is **always stable.** An atom with two **electrons** in its outer shell, is only **stable** when these electrons are the **only two** in the atom. Check the number of electrons in the outer shells of the noble gases (pp. 18–19). (Recall that the noble gases are inert, uncombining atoms.) Do your observations confirm what you have just read about the stability of an atom?

When atoms "join" due to the transfer of electrons, the process is called ionic bonding. An ion is an atom (or group of

Combine Join together.
Stable When matter is stable, its physical and chemical makeup is not easily changed.

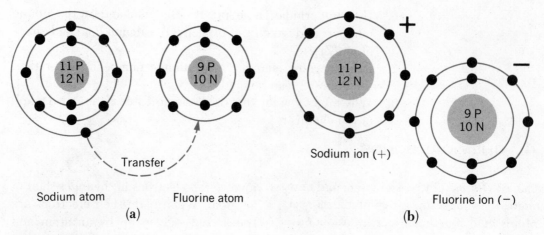

Transfer

Sodium atom   Fluorine atom

(a)

Sodium ion (+)

Fluorine ion (−)

(b)

FIGURE 4-1

atoms acting as a unit) that does not have the same number of protons as electrons. For example, the atoms of sodium and fluorine can combine to form sodium fluoride (Fɪɢ. 4-1a). Sodium has one electron in its outer shell. When the electron transfers from the sodium atom to the fluorine atom, both atoms have eight electrons in their outer shells. Note that the sodium atom no longer has three shells. Its second shell has now become the outer shell.

When one electron of the sodium atom transferred to the fluorine atom, the so-

dium atom became a *positive ion* (Fɪɢ. 4-1b). The sodium is considered *positive* because it has *more protons than electrons*, and protons have a positive (+) electrical charge.

When the fluorine atom received the electron from the sodium atom, it became a negative ion. The fluorine atom is considered *negative* because it has *more electrons than protons*, and electrons have a negative (−) electrical charge.

Negative ions are attracted to positive ions, just as any negative electrical charge is attracted to a positive charge. This at-

FIGURE 4-2   Two fluorine atoms combine to form a fluorine molecule.

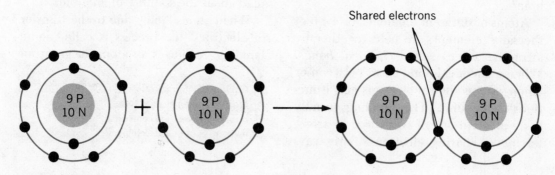

Shared electrons

traction "bonds" the sodium ion and fluorine ion together, and a new substance, sodium fluoride, forms. **When an electron is, or electrons are, at the same time attracted to two nuclei, a chemical bond exists.**

When atoms "join" due to the *sharing* of electrons, the process is called *covalent* bonding. **In covalent bonding the outer shells of the atoms overlap and the atoms share the same electron or electrons.** Two fluorine atoms combine by sharing electrons (Fig. 4-2). Count the number of electrons in each outer shell *before* they combine. *After* they combine you will see that each outer shell has *eight*. The two fluorine *atoms* have bonded together and formed a *molecule* of fluorine.

Broadly speaking, molecules are chemically combined atoms that act as a unit. Strictly speaking, a molecule is made up of two or more atoms that are covalently bonded by the sharing of electrons. Molecules may be simple such as the fluorine molecule. They may also be made up of a number of atoms as in the propane molecule. **This electron dot model shows only the shared atoms of carbon and hydrogen** (Fig. 4-3a). Electron dot models or structural models are used when there are a number of atoms in a molecule (Fig. 4-3b). These formulas are easier to understand. They show only the electrons involved in bonding or the bonds. More than one pair of electrons may be shared between adjacent carbon atoms (Fig. 4-3c). When two pairs of electrons are shared, the bond is called a *double bond*. When three pairs of electrons are shared, the bond is called a *triple bond*. Chemists can show double and triple bonds by using an electron dot or a structural model.

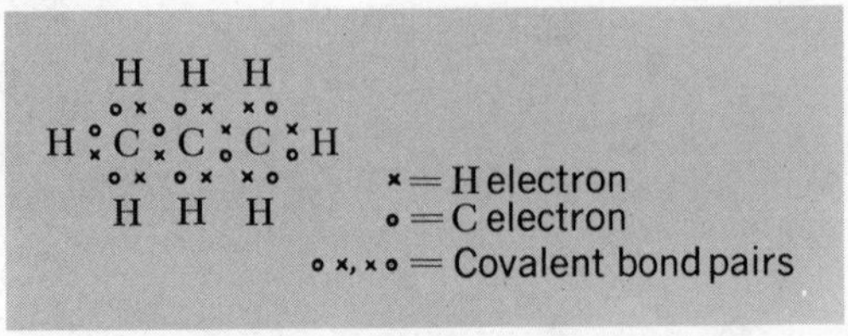

(a)

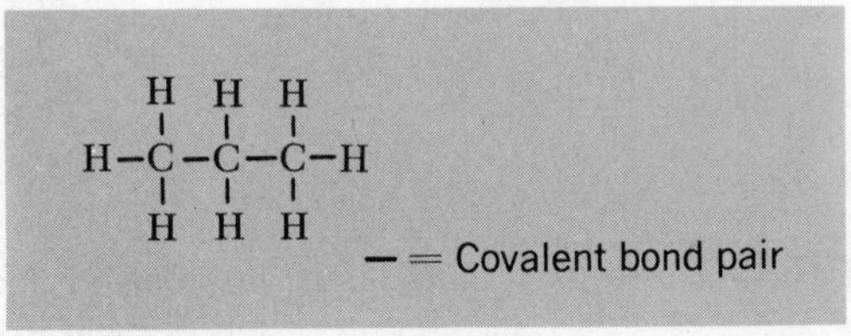

(b)

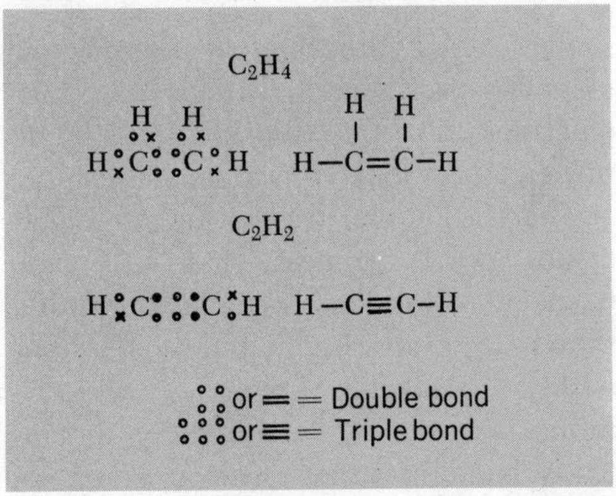

(c)

FIGURE 4-3 (a) Electron dot model of propane. (b) Structural model of propane. In ethene (C_2H_4) two pairs of electrons are shared by the two carbon atoms. In ethyne (C_2H_2) three pairs of electrons are shared by the two carbon atoms.

A water molecule is made up of two atoms of hydrogen bonded to one atom of oxygen (Fig. 4-4). Count the number of electrons in the outer shell of the oxygen atom after it has been bonded. You should find that the outer shell has eight electrons and is therefore stable. The outer

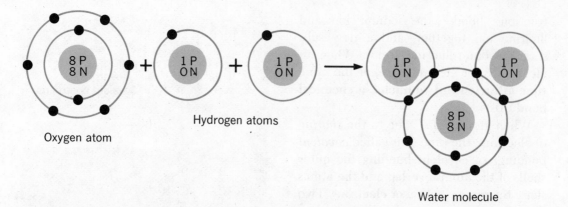

Oxygen atom

Hydrogen atoms

Water molecule

FIGURE 4-4 Compare the number of electrons in the outer shell of the oxygen atom and in the outer shell of each hydrogen atom with the number after bonding to form the water molecule.

shell of each hydrogen atom contains *two* electrons after bonding with oxygen. Two electrons make this atom stable. Why do *two* electrons make this atom stable?

There is another way of showing what a substance is made of. That is to write its chemical formula. **A chemical formula shows the kinds, and number of atoms within a molecule.** (Note: In this definition we refer to combinations of ionically bonding atoms, such as NaCl, as "molecules" even though they are not technically "molecules.")

✺ You have seen that certain atoms can bond with other atoms by means of ionic bonds. The bonding does not take place in a single one-to-one combination (to form a single NaCl unit), but in great arrays, forming the familiar sodium chloride crystal ("table salt"). It is convenient, however, to refer to the simple combination NaCl as the equivalent of a

single *molecule*, although it would be better to give it a different name.

You have seen that a molecule of water contains two atoms of hydrogen and one atom of oxygen. Its chemical formula is written H_2O, and pronounced "H-two-O." H_2 indicates that there are two atoms of hydrogen. O indicates that there is one atom of oxygen in the molecule. If there is only *one* atom of a certain element within a molecule, there is no small number to the right of the symbol. H_2O is written rather than H_2O_1 — although they would both mean the same thing.

The small numbers in chemical formulas are called *subscripts*. Subscripts are always whole numbers rather than fractions. A fractional subscript would indicate that only a *part* of one atom is combining with another. In chemical reactions that take place under normal conditions, that is impossible. How would you write the chemical formula for propane (FIG. 4-3a)?

Some of the laws that determine how elements combine are complex. A few

| *Array* An orderly arrangement. |

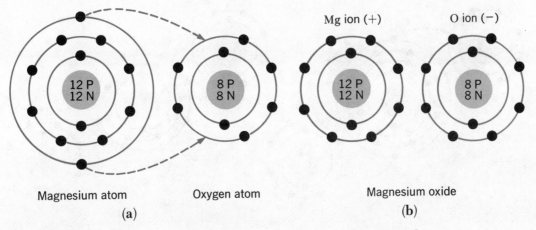

Mg ion (+)   O ion (−)

Magnesium atom   Oxygen atom   Magnesium oxide

(a)   (b)

FIGURE 4-5   Magnesium and oxygen combine to form magnesium oxide.

simple — but very important — generalizations as to how certain families of elements combine can be made by consulting the Periodic Table.

A very common type of combination is the bonding of a *metal* from the left side of the Periodic Table with a *nonmetal* from the right side (pp. 18–19). The elements on the left side of the Periodic Table have very few electrons in the outer shells. They tend to *lose* electrons when combining. Those on the right side have more electrons in the outer shells. They tend to *gain* electrons to complete the outer shells and become stable.

Magnesium oxide is a combination of the elements magnesium (Mg) and oxygen (O). Magnesium is located in Group IIA of the Periodic Table and therefore has two electrons in its outer shell. Oxygen, in Group VIA, has six electrons in its outer shell. When the two atoms combine, magnesium gives up the two electrons in its outer shell to the oxygen atom (FIG. 4-5a,b). Now each atom has eight electrons in its outer shell and is stable.

The magnesium atom becomes a positive ion since it now has more *protons* than electrons. The oxygen atom becomes a *negative ion* since it now has more *electrons* than protons. The two ions are held together by ionic bonding. The chemical formula is written MgO. It shows that there is one magnesium atom for every one oxygen atom in magnesium oxide.

Calcium fluoride is a combination of *one* calcium atom for every *two* fluorine atoms (CaF_2). The Periodic Table shows you that calcium is found under Group IIA of the Periodic Table (pp. 18–19). It therefore has *two* electrons in its outer shell. It could give up the two electrons from its outer shell and become stable.

Fluorine, found in Group VIIA, has seven electrons in its outer shell. It would become stable if it had *one* more electron in its outer shell. Since calcium has *two* electrons it could give away, it will transfer *one* electron to *each* fluorine atom. The three atoms become ions and are held together by ionic bonding (FIG. 4-6a,b).

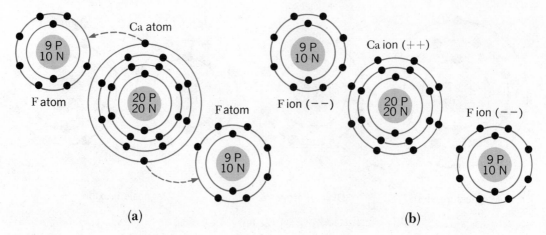

(a)   (b)

FIGURE 4-6 In the formation of the ionic compound calcium fluoride (b), electrons are transferred from the calcium atom to the fluorine atoms (a).

✻ The *outermost shell* (orbit) of an uncombined atom is called its *valence shell*. The electrons in the valence shell are called *valence electrons*. **In most atoms, the valence electrons are the only electrons that enter into a chemical reaction.** Each atom has a valence. **The valence of** an atom is equal to the number of electrons that it can transfer or share when bonding with other atoms. The valence has a positive (+) sign if it is most likely to *lose* electrons. The valence has a negative (−) sign if it is most likely to *gain* electrons.

BE CURIOUS 4-1:   **Find out if you can construct electron dot models.**

The substances listed as 1, 2, 3 (NaCl, LiF, BeF$_2$) are ionically bonded. Draw electron dot models that show the transfer of electrons and the ions that result. Tell which of the resulting ions is *negative* or *positive*.

The substances listed as 4, 5, 6 are covalently bonded. Draw electron dot models of each to show the *covalent* bond. Refer to the Periodic Table to find electron arrangements (pp. 18–19).

1. Sodium Chloride (Na Cl)   4. Hydrogen Chloride (H Cl)
2. Lithium Fluoride (Li F)   5. Carbon Dioxide (CO$_2$)
3. Beryllium Fluoride (Be F$_2$)   6. Carbon Tetrachloride (C Cl$_4$)

OBJECTIVE 4
ACCOMPLISHED?
FIND OUT.

1. What is meant by the term chemical bond? When a chemical bond exists among atoms, how do the atoms behave?
2. Describe and name the two ways atoms chemically bond with one another.
3. What is a *positive* ion? What is a *negative* ion?
4. Under what conditions does a chemical bond exist?

5. What family of elements normally do *not* combine with the other elements? Why not?
6. One atom from Group IA of the Periodic Table should combine most easily with one atom from what other group of the Periodic Table? (You may consult pp. 18–19 to answer this question.)
7. What is a molecule?
8. What information do you get from a chemical formula?
9. Why do metals tend to combine with nonmetals?
✧ 10. What is meant by the term valence?

5 | ELEMENTS, COMPOUNDS AND MIXTURES

YOUR OBJECTIVE: To find out how an element, a compound, and a mixture are alike and how they differ; to understand how some compounds are named.

An element is a *simple* substance. **An element cannot be broken down into any simpler substances by ordinary chemical means. An element is made up entirely of one type of atom.** The atoms of an element may be arranged as individual atoms or as molecules. Elements in which the atoms exist as *individual atoms* are represented *only* by their *symbol*. The element helium is made up of individual atoms. So, the symbol He correctly represents the element helium (FIG. 5-1a). Atoms of the element oxygen normally exist in *pairs* (a molecule of oxygen). So, the symbol O_2 correctly represents the element oxygen (FIG. 5-1b).

A compound is made up of chemically bonded atoms of two or more different elements. The bonding between the

FIGURE 5-1

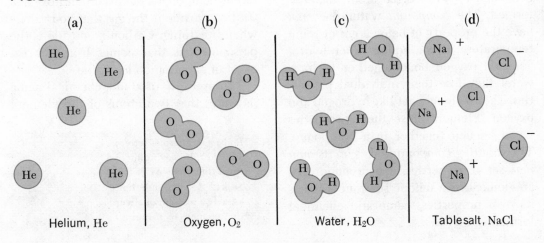

(a) Helium, He (b) Oxygen, O_2 (c) Water, H_2O (d) Table salt, NaCl

atoms can be either *covalent* or *ionic*. In a compound, the different types of atoms are chemically bonded in the *same proportion throughout the compound*. Thus, a compound can be represented by a formula.

Water is a compound made up of atoms of hydrogen and oxygen. The atoms are held together by covalent bonding (Fig. 5-1c). The chemical formula for a molecule of water is H_2O. The formula shows that the proportion is *two* hydrogen atoms for every *one* oxygen atom in every molecule of water.

Table salt is a compound made up of atoms of sodium and chlorine. The sodium and chlorine ions are held together by ionic bonding (Fig. 5-1d). Throughout a salt crystal (array), the atoms are in a *proportion* of *one* sodium atom to every *one* chlorine atom. The chemical formula is NaCl.

When elements combine to form compounds they lose their individual properties. The *element* hydrogen is a colorless gas at room temperature. The *element* oxygen is also a colorless gas at room temperature. When two atoms of hydrogen bond to one oxygen atom, *water* is formed. The *compound* water does *not* have the property of being a gas at room temperature. It is a liquid. When hydrogen and oxygen bond together to form water they lose their individual properties. They no longer act like hydrogen and oxygen. Elements *lose* their properties when bonded together in a compound. The resulting *compound* takes on its *own new set* of properties. A compound, like an element, is a unique substance. It has its own properties, name and chemical formula.

Rust is a *common name* for a compound of iron and oxygen. *Iron oxide* is the *chemical name* for this combination. Chemical names of compounds usually give you a clue as to what elements the compounds are made up of. Many times the chemical name even tells you the proportion in which those elements are found in the compound. **Chemical names of compounds usually are combinations of the names of the elements they contain plus prefixes and suffixes.** Prefixes and suffixes may indicate how many atoms there are or how the compounds formed. A few examples will help you learn how to read and write chemical names of some *simple* compounds.

When chemists write the name of a compound made from a metal and a nonmetal, they place the name of the metal first. When oxygen bonds with iron the chemical name is iron oxide — not oxygen ironide. Other examples are lead sulfide and sodium chloride.

Sometimes a *prefix* is needed to identify one compound from another made of the same elements, but in *different proportions*. For example, there are two different compounds of carbon and oxygen. Carbon *di*oxide is the gas that you exhale when breathing. Carbon *mon*oxide is the poisonous gas that comes from the exhaust of an automobile engine.

The prefix *di-* used in a chemical name indicates that two atoms of the element

Prefix A group of letters added to the beginning of a word that changes its meaning.
Suffix A group of letters added to the end of a word that changes its meaning.

TABLE 5-1	
Chemical	Chemical Formula
Carbon tetrachloride	CCl_4
Tin dioxide	SnO_2
Calcium oxide	CaO
Nitrogen pentoxide	N_2O_5
Dinitrogen pentasulfide	N_2S_5
Sulfur trioxide	SO_3
Tellurium monoxide	TeO

are present in the compound. Carbon dioxide (CO_2) is a compound made up of one atom of carbon and two of oxygen. The prefix *mon-* means *one*. The compound made up of *one* atom of carbon and *one* of oxygen is called carbon *mon*oxide (CO). The prefix *tri-* refers to *three; tetra-* to four; *penta-* to five (TABLE 5-1).

The suffix *-ide* is frequently added to the name of the nonmetal in a two-element compound. For example *ox* of oxygen plus *ide* form oxide. *Chlor* of chlorine plus *ide* forms chloride. You may wonder why chemists go to the trouble of forming new names. It is because a new "group" identification name tells the chemist something. In the case of the *ide* compounds, the *ide* signifies a group of compounds that are related. The compounds form in similar ways and have some properties in common. It is easier for a chemist to remember properties common to a *group* of compounds than to remember the properties of many individual compounds. Knowing the group properties of *ides* and the properties of the metallic element in a compound the chemist can make a good guess as to the properties of each individual *ide* com-

pound. If you go on to study chemistry, you will become acquainted with the meanings of other prefixes and suffixes.

Laboratory investigations of matter are investigations of physical changes or chemical changes. Investigations of physical change often involve working with tools or machines. Investigations of chemical change involve chemicals. When working with *mechanical* things many of the dangers are very obvious. You can *see* that you must protect yourself from such things as turning gears, rotating fans, and sharp objects. When working with *chemicals* however, most of the dangers are *hidden*.

It is the hidden danger that often causes accidents and injury. Unless you have a great knowledge of chemistry it is impossible to predict what will happen when working with various chemicals. Some chemicals cause severe burns to the skin or eyes. The fumes of many chemicals are dangerous. Certain mixtures of chemicals are very explosive.

There is no *one* set of safety rules that can cover all work with chemicals. Therefore, *never* conduct a chemical experiment without the knowledge and supervision of your instructor. A few general safety rules are listed for your protection. The following suggestions can prevent unfortunate accidents.

1. Know where all emergency equipment and supplies are located *and* how to use them. Emergency equipment includes fire extinguishers, fire blankets, and running water.
2. Be sure that there is proper ventilation in the area where you work with chemicals.

Find out if you can tell by identifying properties of individual elements whether the two elements when combined form a compound or a mixture.

Part A

Powdered sulfur
Iron filings
Sheet of paper
Magnifying glass
Stirring rod
Magnet

Before beginning any part of this experiment be sure that you have read the section on laboratory safety.

1. Place about a teaspoon of powdered sulfur in a pile on a sheet of paper. Next to it, place the same amount of iron filings. Examine each pile with a magnifying glass. Describe each of the elements.
2. Test the sulfur and the iron with a magnet to see if either are attracted to the magnet. What were the results?
3. With a stirring rod or spoon mix the sulfur and iron together very thoroughly. Examine the mixture with a magnifying glass. Have the characteristics of the iron or sulfur changed?
4. Try to separate the iron from the sulfur by using a magnet. Can it be done? Can you think of any other ways that the two elements might be separated? Save the iron and sulfur for Part B of this investigation.

Part B

Test tube
Test tube tongs
Bunsen burner
Towel
Hammer
Stirring rod or
*   spoon*

Place the mixture of iron and sulfur from Part A into a test tube. Light the Bunsen burner. Heat the test tube until the contents are red hot (see the figure at left). When you heat a test tube and contents over a flame, gently shake the test tube as you heat it. Be sure the test tube opening is not directed toward your face or anyone else's face. Then remove it from the flame. Place the test tube into a pyrex beaker (or cup) to cool. After it has cooled, wrap the test tube in a towel and break it with a hammer. With a stirring rod (or spoon) carefully remove some of the contents from the broken test tube.

1. Examine the material with a magnifying glass. Describe the material. Can you identify the elements iron and the sulfur in the material? Explain your answer.
2. Try to separate the iron filings from the sulfur by using a magnet. What happens? Explain.
3. On the basis of what you have observed do you think the heated iron and sulfur have gone through a physical or chemical change? What do you think the ''heated'' material is? Explain your answers.

3. Avoid clutter on the lab working surface. Do not place chemicals near the edge of the table where they can be easily spilled.

4. Do not taste or smell chemicals unless directed to do so. When directed, check with your teacher as to the proper way to detect the taste and odor of chemicals.

5. Wear safety goggles when working in the laboratory. Injuries to the eyes can be very serious.

6. When heating chemicals in a test tube, do not point the open end of a test tube at anyone, including yourself. Do not look directly into the opening of a test tube while working with chemicals.

7. Keep flames and sparks away from possible explosive or burnable material. Keep your hair away from any flame.

8. Use heat-resistant glassware when heating chemicals. Bring the glassware into the heat source slowly when starting the heating.

9. When inserting glass tubing (or thermometers) into the holes of stoppers be careful of breakage of the glass and injury to the hand. Lubricate the tubing with water and with a towel grasp the part of the tubing nearest the stopper. Slowly *twist* the tubing as you push it into the stopper. (FIG. 5-2).

In a mixture the individual materials do not lose their characteristics. In Part A of *Be Curious 5-1* you found that iron and sulfur *kept* their own characteristics after being mixed together. The elements

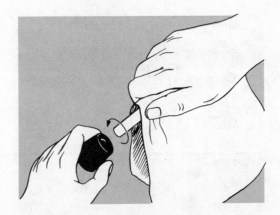

FIGURE 5-2

were not chemically bonded and could be easily separated. That combination of iron and sulfur was *not* a compound. It was a *mixture*.

In Part B, the sulfur and iron chemically bond to form the *compound* iron sulfide (FeS). The chemical formula should tell you that *one* atom of iron combined with *one* atom of sulfur. If there were more iron atoms than sulfur atoms placed into the test tube, you may have been able to detect some uncombined iron remaining in the test tube after heating. The reverse would be true if there were more sulfur atoms present than iron atoms.

A combination of various elements or compounds in which all elements are not chemically bonded together is not a *pure* substance. **The individual materials that make up the mixture may be mixed together in any proportions.**

Air is an example of a mixture. It is a mixture of many different gases including nitrogen, oxygen, carbon dioxide, carbon monoxide and water vapor. These gases

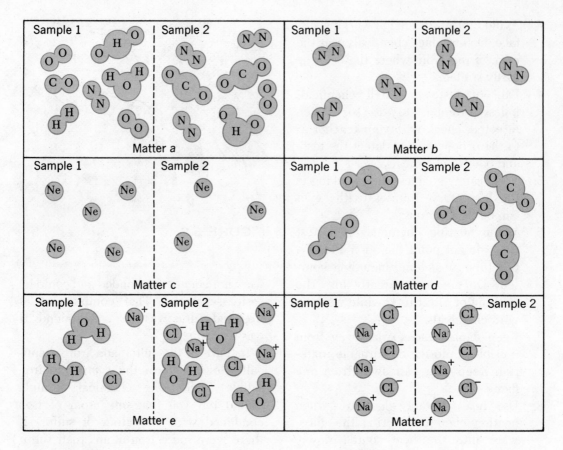

FIGURE 5-3

BE CURIOUS 5-2:   **Can you tell by looking at models of atoms that make up matter whether the matter is an element, compound or mixture?**

Part A

Examine the models that show the arrangements and proportions of atoms which make up the six different kinds of matter identified as Matter a, b, c, d, e, f (FIG. 5-3). Note that you are shown two samples of each kind of matter. How will two samples aid you in identifying a mixture? Tell whether each kind of matter represented is an *element, compound* or *mixture.* Give the reasons for each of your answers.

Part B

If you can, give the common or chemical name for each of the kinds of matter shown.

are found in various proportions in air samples taken from different places. The amount of oxygen found in the air of a burning building is certainly less than that found in the air outdoors. The amount of carbon monoxide found in the air along a busy city street is usually greater than that found in a rural area.

Elements and compounds in a mixture each have the same characteristics that they would have if not mixed together. Oxygen supports combustion in its pure form or when it is in the mixture of air. Carbon monoxide is poisonous when in its pure form and when it is part of the air mixture. Luckily, there is very, very little carbon monoxide in the air you normally breathe so it does not poison you.

OBJECTIVE 5
ACCOMPLISHED?
FIND OUT.

1. What is an element?
2. What is a compound? How are atoms in a compound held together?
✴ 3. How is an element like a compound?
4. What is the procedure for naming a compound made of a metal and a nonmetal?
5. What do the prefixes *mon, di,* and *penta* indicate when used as part of a name in a chemical compound?
6. If you do not follow the safety precautions listed on pages 31 and 33, you would injure yourself or others. Which precautions will protect your eyes and face from injury?
7. What is a mixture? How does it differ from a compound?
✴ 8. What is the advantage of naming compounds the way chemists do?

6 | SOLUTIONS AND SUSPENSIONS

YOUR OBJECTIVE: To find out what a solution and a suspension are; to describe each; to investigate ways of separating materials in each.

A mixture of salt and water is called a solution. **A solution is a mixture of atoms, ions, or molecules in which the atoms, ions or molecules are in nearly the same proportions throughout.** Ocean water is a solution. If you took a cup of ocean water from the Atlantic Ocean it would contain nearly the same amount of salt as a cup taken from the Pacific Ocean.

In a solution one (or more than one) material is dissolved in another. **The material that does the dissolving is called the solvent. The material that is dissolved is the solute.** In a salt water solution, water is the *solvent* and salt is the *solute.*

Not all materials can be dissolved by a single solvent. Salt will dissolve in water, but grease will not. Salt is then said to be *soluble* in water, while grease is said to be *insoluble. Water is the most common solvent.* It is capable of dissolving so many different materials that it is sometimes called the *universal solvent.*

BE CURIOUS 6-1: Find out if the solvent and solute in a solution lose or keep their individual properties.

Salt
Spoon
Glass
Water
Funnel
Filter paper
Quart jar or bottle
Saucer

Filter paper

Place about a teaspoon of table salt in a glass of water. Stir thoroughly. After stirring, place the container of salt water on a table top and do not disturb it for a few minutes. What do you observe?

Pour the mixture of salt and water through a filter (see figure at left). What do you see on the filter paper? Was the salt removed from the water by the filter?

Take a small amount of the water that passed through the filter and pour it into a saucer. Place the saucer in a part of the room where it can remain overnight. Inspect the saucer the next morning. What do you observe? Wet your finger and touch whatever is in the saucer. Place your finger on the tip of your tongue and taste. Immediately rinse your mouth with water. What can you conclude from your observations? What evidence do you have that salt and water formed a solution? What evidence do you have that salt or water kept their individual properties?

Table salt dissolves when mixed with water. You observed this in *Be Curious 6-1*. The salt seemed to become a permanent part of the liquid. It did not settle to the bottom of the water after you stopped stirring it. It could not even be removed by filtering.

Most people think of a solution as always being made up of a *solid* solute dissolved by a *liquid* solvent. This is not true of all solutions.

Solutions can be made up of various combinations of solid, liquid or gas. Salt in water is an example of a solid dissolved in a liquid. Soda water is an example of a gas dissolved in a liquid (Fig. 6-1). Ink and water form a liquid in liquid solution. Brass is a solid solution of copper and zinc, both solids. How do you think one solid can be made to dissolve in another?

The speed at which a solid dissolves in a liquid is called its *rate of solution*. A solid will dissolve faster in a *hot* liquid than in the same *cool* liquid. A solid that is *broken* into many small pieces will dissolve faster than if it were left in one large piece. *Stirring* a mixture will cause the solid to go into solution faster.

If you wish to dissolve a solid in a liquid solvent as quickly as possible you can do three things: *heat* the liquid; *crush* the solid into a fine powder; *stir or shake* when mixing the powder and the liquid.

A concentrated solution is one in which there is a large amount of solute. A dilute solution is one in which there is a small amount of solute.

A solution can be made more concentrated in two ways. **More solute can be**

Concentrated In reference to solutions, to have much dissolved solute. Strong.
Dilute In reference to solutions, to have little dissolved solute. Weak.

FIGURE 6-1 The bubbles you see are bubbles of carbon dioxide escaping from the soda water solution. *(Courtesy of the Seven-Up Company and 7UP.)*

added to make a more concentrated solution. **The amount of solvent can be reduced to make the solution more concentrated.** For example, a solution of salt water could be made more concentrated by adding more solute (salt). It could also be made more concentrated by allowing some of the solvent (water) to evaporate.

A solution can be made more *dilute* in the opposite manner. **More solvent could be added to make a more dilute solution. Also, the amount of solute can be reduced to dilute the solution.**

A solution can be made more concentrated by adding more solute. But if you

---

Evaporate To change from a liquid to a gas.

---

kept adding more salt to water, you would find that the solution would reach a point at which no more salt could be dissolved. **The point at which a solution can dissolve no additional solute is called the saturation point.**

If you keep adding solute to a saturated solution, you will find that it will not dissolve. The additional solute does not become part of the solution. If you go beyond the saturation point by adding too much salt to a salt-water solution, you will find that the additional salt will settle to the bottom of the container. This additional salt can be removed by filtering, since it is not a part of the solution.

The saturation point for a solute in a solvent can be changed by changing the temperature of the solvent. When a liquid solvent is heated, generally speaking, it is able to dissolve *more* solid solute before the solution becomes saturated. Warm water can hold more dissolved salt than cold water can.

You found in *Be Curious 6-1* that a solute could *not* be separated from a solution by filtering. Since a solution is a *mixture* there should be a simple way to separate the solute and the solvent. One way is to allow a liquid solvent to evaporate and leave the solid solute behind.

If you wanted to remove the salt from salt water you could use the *evaporation* method. A shallow container of salt water could be placed in an area where the water is allowed to evaporate. Within a short period of time the water could have evaporated into the air. The solid salt would be left behind in the container.

Suppose that you were very thirsty and all that you had to drink was salt water. Would the evaporation method be good

BE CURIOUS 6-2: **Find out if soil is soluble in water.**

Soil, spoon, glass
Water, funnel
Filter paper
Quart jar or bottle

Place about a teaspoon of soil in a cup of water. Stir thoroughly. What do you observe? Place the cup on a table top and do not disturb it for a few minutes. What do you observe? Did the soil and water form a solution? What is your evidence?

Stir the mixture of soil and water again. Immediately pour the mixture through a filter. Observe the filter paper and the water. Was the soil removed by the filter?

to use for removing the salt from the water that you wanted to drink? Using this method you would have saved the solute (salt) but lost the solvent (water). There is a method of separating the solute and the solvent in which neither is lost. This method is called *distillation.*

In distillation a liquid solution is heated to the boiling point of the solvent, causing the solvent to evaporate. Instead of escaping into the air, the resulting gas passes through a glass tube where it is cooled and turned back into a liquid. The

pure solvent is collected as it drips from the tube. This method of distillation can also be used to separate a solution made up of two or more liquids — since various liquids have different boiling points, one will evaporate before the other.

You found that the soil in *Be Curious 6-2* did *not* dissolve in water. The soil could be mixed with the water, but it settled to the bottom when the mixing stopped. The soil was "held up" (suspended) in the water because the stirred water was in motion. **When materials are**

FIGURE 6-2 (a) In a freshly-prepared mixture small particles of rock and soil are suspended in water. (b) An hour later most of the particles have settled to the bottom. (*Courtesy of Fundamental Photographs from the Granger Collection.*)

(a)

(b)

"held up" in a fluid (a liquid or gas), such as soil in water and dust particles in air, the mixture is called a suspension (FIG. 6-2a).

Suspensions separate very easily. Suspended material "held up" by motion of the fluid settles to the bottom after motion has stopped (FIG. 6-2b). Material in a suspension can also be separated from the fluid by filtering.

Small particles are usually held in suspensions more easily than large ones. A river may be able to carry along small particles of clay in suspension, but not larger grains of sand or still larger pebbles. If a mixture of different sized particles are suspended by a liquid the larger particles will settle to the bottom first.

OBJECTIVE 6
ACCOMPLISHED?
FIND OUT.

1. What are the main differences between solutions and suspensions?
2. Suppose that you were to dissolve a solid solute in a liquid solvent. List the three things that you could do to increase the *rate of* solution.
3. How can a solution be made more concentrated?
4. How can a solution be made more dilute?
5. What is meant by the term saturation point?
6. What is a suspension?
✴7. Describe one way a solution made up of two liquids can be separated.

7 | ACIDS, BASES AND SALTS

YOUR OBJECTIVE: To find out what acids, bases, and salts are; to describe the characteristics of each; to investigate the relationship of acids, bases and salts.

Matter is made up of elements in their pure form or of combinations of elements. Combinations of elements are either mixtures or compounds. Chemists group compounds in various *classes* according to their chemical properties. Three important classes are called acids, bases, and salts. Why can't chemists group mixtures in various classes according to their chemical properties?

In the acid class there are weak acids and strong acids. You are probably familiar with some of the *weaker acids* — although you may not have identified them as acids. *Citric acid*, found in such fruits as oranges, lemons and grapefruits, gives the fruits their sour taste. *Acetic acid* is found in vinegar. Sour milk contains *lactic acid*, and dissolved aspirin contains *acetylsalicylic* (ə•'sēt•°l•ˌsal•ə•ˌsil•ik) *acid*.

Acids such as hydrochloric acid and sulfuric acid are considered very strong acids. Strong acids can be diluted many times with water. Even when diluted they must be handled with great care since they can cause very severe burns. In a chemistry laboratory three strong acids you will always find are *sulfuric, hydro-*

chloric and *nitric* acids. You will see two, or more, different kinds of bottles of each acid. In each bottle the ratio of acid to water will differ. For instance, in most chemistry laboratories you will find a bottle of *concentrated* sulfuric and a bottle of *dilute* sulfuric acid.

In TABLE 7-1 are listed four common acids and their formulas. What element do *all* of the acids contain? What element do *most* of the acids contain? Note the *other* elements found in these acids. Where are they located in the Periodic Table (pp. 18–19)?

All of the acids contain *hydrogen* and many of them contain *oxygen*. The other element that is combined with the hydrogen and oxygen can usually be found toward the right-hand side of the Periodic Table. Elements on the right-hand side are all *nonmetals*. The *nonmetals* found in the four acids listed in TABLE 7-1 are chlorine, nitrogen, sulfur and carbon.

One property of acids is that they have a *sour* taste. The sour taste of vinegar, lemons and sour milk is due to the acids found in each of them. Since some acids can cause severe burns it is *not* a good idea to taste materials to see if they might be acids.

A chemically treated paper called *litmus paper* is a safe way to tell whether a substance contains an acid. There are two kinds of litmus paper, blue and red. **If *blue* litmus paper is placed in an acid it will turn from *blue* to *red*.** If an acid is not present the litmus paper will remain blue. A material (such as litmus paper) that *indicates* whether an acid or base is present is called an *indicator*.

One of the most useful acids is **sulfuric acid.** Along with dilute hydrochloric acid,

TABLE 7-1	
Acid	*Chemical Formula*
Hydrochloric acid	HCl
Nitric acid	HNO_3
Sulfuric acid	H_2SO_4
Acetic acid	$HC_2H_3O_2$

a very dilute sulfuric acid is one of the acids which in the human body digests food. In a much more concentrated form it is the acid used in automobile batteries. More than one auto mechanic has sat down to rest on an automobile battery only to find that holes developed in the seat of his pants. That is because some acid was present on the surface of the battery, and "ate" into his pants. Metals that are to be painted or coated are often treated with acid first. In industry the acid used is usually sulfuric acid.

Nitric acid is used in the manufacturing of fertilizers. And, it is also used in the production of explosives, including nitroglycerin and dynamite.

How do bases differ from acids? Study TABLE 7-2. It lists four common bases and their formulas. What two elements do *all* of the bases contain? Note the *other* elements found in these bases. Where are they located in the Periodic Table (pp. 18–19)?

TABLE 7-2	
Base	*Chemical Formula*
Calcium hydroxide	$Ca(OH)_2$
Sodium hydroxide	$Na\ OH$
Magnesium hydroxide	$Mg(OH)_2$
Aluminum hydroxide	$Al(OH)_3$

All of the bases contain a combination of oxygen and hydrogen. This combination of oxygen and hydrogen is called a *hydroxide (OH)*. The chemical formula for calcium hydroxide shows that there are two units of hydroxide (two atoms of oxygen *and* two of hydrogen) for every one atom of calcium.

In addition to the hydroxide, bases usually contain an element from the left-hand side of the Periodic Table. These elements are *metals*. The metals found in the four bases listed above are calcium, sodium, magnesium, and aluminum.

Chemical names for bases usually consist of the name of the metal followed by "hydroxide." You are probably familiar with the *common names* of some bases. Milk of magnesia is the common name of magnesium hydroxide. Lye is the common name for sodium hydroxide, and ammonia water for ammonium hydroxide.

Bases have a *bitter* taste. They also have a *slippery* feel. Some bases can burn the skin, since bases react with organic matter. Tasting and touching are *not* good ways to test for a base in a substance. One of the easiest ways to test for a base is to use litmus paper. **If *red* litmus paper is placed in a base, the color of the paper will turn from *red* to *blue*.** If a base is not present, the color of the litmus paper will remain red.

In many ways, bases are the opposite of acids. Bases contain *metals*. Acids contain *nonmetals*. Bases turn *red* litmus paper to *blue*. Acids turn *blue* litmus paper to *red*. Bases taste *bitter*. Acids taste *sour*.

When a base is added to an acid the mixture becomes less like an acid. Depending on how much base is added

TABLE 7-3	
Salt	Chemical Formula
Sodium Chloride	NaCl
Calcium Chloride	$CaCl_2$
Trisodium phosphate	Na_3PO_4
Potassium nitrate	KNO_3

the mixture remains acidic, becomes basic or neutral. **A neutral material is one that will *not* change the color of either red or blue litmus paper.** If you add a proper amount of a base to an acid, the base will completely *neutralize* the acid. The neutralized mixture will not change the color of either red or blue litmus paper. If you then add more of the base to the neutralized mixture, you will get a basic reaction (red to blue) with litmus paper.

If an acid and a base combine in the proper amounts, they produce a compound called a salt. This word equation that follows states that an *acid* plus a *base* produces a *salt* plus *water*.

acid + base ⟶ salt + water

TABLE 7-3 lists a few common salts. Note that some of the salts contain oxygen while others do not. Every salt contains one element from the left-hand side of the Periodic Table — a *metal;* and, one element from the right-hand side of the table — a *nonmetal*.

There are many different salts. **The properties of salts vary. Many salts do not**

Equation In chemistry a symbolic representation of a chemical reaction.

change the color of red or blue litmus paper. Would you call such salts acidic, basic or neutral? The taste of salts can range from bitter to "salty." Salts come in different colors, but many are white.

The best known salt is table salt or sodium chloride (NaCl). Sodium chloride has many uses. The best known use is the salting of foods. Not only does this use improve the flavor of foods, but the addi- tion of salt to food provides salt needed by the body for proper operation of some bodily functions. Sodium chloride in the form of rock salt is commonly used to help melt ice and snow on sidewalks and roads. Sodium chloride is also used in the preparation of many chemicals that are used in industry and research.

The chemical equation shown on p. 91 states that *hydrochloric acid* (HCl) plus

BE CURIOUS 7-1:

Identify some common acids and bases and what they are used for.

Part A

Would an *acid* or a *base* be the better substance to use for each of the following purposes? Give reasons for your answers.
a. Cleaning a "clogged" sink drain
b. Removing the dull surface of a metal to expose the shiny metal under it
c. Neutralizing excessive acid in the stomach
d. Cleaning grease from the inside of an oven

When you have completed Part B and Part C of this investigation, review the responses you made to a–d. Will you change any responses? Compare your conclusions with your classmates' findings.

Part B

6 Saucers
Distilled water
Vinegar, soapy water
Milk of magnesia
Table salt
Aspirin, water
Red and blue litmus
*   paper*

The results from this part of the investigation will help you answer some of the questions in Part A. Place a small amount of each of the following materials in separate saucers: distilled water, vinegar, soapy water, milk of magnesia, table salt in water, and aspirin dissolved in a small amount of water. Predict whether each solution is an *acid*, a *base*, or *neutral*. Test your predictions by touching each solution with red litmus paper. Repeat with blue litmus paper. Which solutions are acid? Which base? Are any of the solutions neutral?

Part C

Study the labels on household products used in your home. Make a list of those that contain acids or bases. List the name of the acid or base together with the commercial name of the product. Also list what each product is used for.

sodium hydroxide (NaOH) produces *sodium chloride* (NaCl) plus *water* (H_2O).

$$HCl + NaOH \longrightarrow NaCl + H_2O$$

When salt is produced in this manner the *nonmetal* from the acid combines with the *metal* from the base. In the example above, the *chlorine* from the acid combined with the *sodium* from the base. In other reactions some hydrogen or oxygen also joins with the metal and nonmetal to form a salt (KNO_3). The remaining

"Shortages" are not unique to the 20th century. Shortages of one thing or another have occurred at many times throughout history. Of course, *what* is in short supply changes as the needs of the population, the abundance of resources, and availability of certain products change. At many times in history salt has been in short supply. Wars have been lost because armies lacked salt.

America has had her battles for salt. In colonial America most salt was imported from England. When the Revolutionary war began, this source of supply was cut off. A shortage of salt resulted. Colonists turned to the oldest known method of salt production — evaporation of sea water by the sun — to obtain salt.

The Continental Congress and the state of Pennsylvania spent about $32,000 — a large sum of money for those days — to set up saltworks on the New Jersey coast at the mouth of the Toms River. A windmill pumped salt water from Barnegat Bay to vats. The heat of the sun evaporated the water from the vats leaving the salt behind. Many similar private saltworks also were constructed along the Jersey coast as huge quantities of salt were needed for the manufacturing of gunpowder, and the flavoring and preserving of food. The British, considering salt manufacture vital, in 1778 sent 135 soldiers to burn the saltworks at Toms River. They succeeded.

This woodcut shows how salt was obtained by evaporation of sea water in the 16th century. Note gates regulate the amount of water entering the main vat and each smaller vat.

Again, in the War of 1812, salt imported from England was hard to obtain and cost $5.00 a bushel. Because of this, commercial salt manufacture was begun at Syracuse, New York. During the Civil War, Syracuse production freed the North of all salt worries, but by 1863 the Southerners couldn't buy salt at any price. Salt factories in Virginia and deposits of salt along the Louisiana gulf coast were lost by the South to the North. Shortage of salt in the South influenced the duration of the war.

hydrogen and oxygen unite to form water.

Sodium chloride can be produced in the laboratory from the reaction of hydrochloric acid with sodium hydroxide. But, most salt is obtained from other sources. Sodium chloride is commonly found in nature. A plentiful source is salt water. At times the salt is taken directly from the oceans or a salt lake. The salt water is pumped into a shallow basin and the water allowed to evaporate. The salt that is left behind can then be scraped up and removed from the basin.

Millions of years ago, much salt was deposited in North America. The salt was deposited by inland seas that covered large portions of the continent. When the water from the seas evaporated it left the salt behind. Some of this salt can be found as part of the earth's surface crust. The area near the Great Salt Lake of Utah has much salt on the surface of the crust.

Most of the salt deposited by the ancient inland seas has been buried by other earth materials. This salt can be recovered by mining operations or by sinking salt wells. Mining salt is similar to mining coal. The salt may be blasted loose with explosives and brought to the surface.

Removing salt from below the earth's surface by means of wells is a little more complicated. A well is sunk. Water is pumped *down* the well into the salt deposit. The water dissolves the salt. The salt-water solution is then brought up to the surface by pumps and stored in shallow basins where the water evaporates, leaving the salt behind.

BE CURIOUS 7-2: **See if you can obtain salt crystals from a salt water solution.**

Glass
Water
Table salt
Spoon
Microscope slide
Microscope

Prepare a saturated solution of table salt in water by stirring table salt into a glass until no more can dissolve. Place a few drops of the solution on a microscope slide. As the water evaporates, observe the formation of the salt crystals (see figure at right).

Describe the salt crystals. How can you determine if the crystals are sodium chloride? See if you can obtain sodium chloride from a salt water solution.

OBJECTIVE 7 ACCOMPLISHED? FIND OUT.

1. What type of elements are found in acids? in bases? in salts?
2. Compare the properties of acids and bases.
3. What is an indicator?
4. How can an *acid* be neutralized? How can a *base* be neutralized?
5. What types of material are obtained when an acid and a base are combined?
✳6. In *Be Curious 7-2*, why were you directed to prepare a saturated solution of table salt and water instead of dissolving only a small amount of the salt in water?

Matter is anything that has weight and takes up space. Matter is either organic (contains carbon) or inorganic (does not contain carbon). Chemistry is the science that deals with the composition of matter and changes in the composition of matter.

All matter is made up of very small "building blocks" called atoms. Atoms are mostly empty space. Atoms are very tiny. Atoms are represented by models. A model is anything that represents something else. The three important particles that make up an atom are electrons, protons, and neutrons. Protons and neutrons are found in the nucleus of an atom. Protons have a positive charge. Neutrons have no charge. Electrons have a negative charge. Electrons orbit the nucleus of an atom. The number of electrons in an atom equals the number of protons. Atoms of the *same* element that vary as to the number of neutrons in the nucleus are called isotopes.

The atomic mass unit (a.m.u.) is used to refer to the mass of an atom. The atomic weight of an atom is equal to the total number of protons and neutrons in the nucleus. Each atom has its own atomic number, which is equal to the number of protons in an atom of that element.

An element is a material made up entirely of one kind of atom. Chemists use chemical symbols to represent elements. The symbol stands for one atom of the element. Each element has its own unique structure and chemical and physical properties. Elements with similar properties are grouped together in families.

The Periodic Table lists all known elements in order of increasing atomic number. Elements in the same family appear in the same vertical column of the table. Elements in the same vertical column have the same number of electrons in their outer shell. Elements in the same horizontal row of the table are said to belong to the same period. Elements of the same period have the same number of electron shells, or orbits, surrounding the nucleus.

Atoms of the same element or of different elements are held together by attractive forces called chemical bonds. There are two types of chemical bonds — covalent and ionic. In covalent bonding, electrons are shared between atoms. In ionic bonding, electrons are transferred from one atom to another. When electrons are transferred from one atom to another each atom involved has an electric charge — one a positive charge, the other a negative charge. Charged atoms are called ions. An ion is an atom that does not have the same number of protons as electrons.

Molecules are chemically combined atoms that act as a unit. A common type of combination is the bonding of a metal with a nonmetal. The ability of an atom to combine is called its valence. The valence of an atom is equal to the number of electrons that it can transfer or share when bonding with other atoms. A chemical formula shows the kinds and number of atoms within a molecule.

A compound is made up of chemically bonded atoms of two or more different elements. Elements in a compound combine in definite porportions. Each compound has unique properties. Chemical names of compounds usually are combinations of the names of the elements they contain plus prefixes and suffixes.

The elements and/or compounds that make up a mixture are not chemically bonded with one another. Elements and compounds in a mixture each keep their own unique properties. Elements and/or compounds in a mixture may be mixed together in any proportions.

A solution is a mixture of atoms, ions or molecules in which the atoms, ions or molecules are in nearly the same proportion throughout. Solutions can be mixtures of solids, liquids or gases. In a solution one (or more than one) material is dissolved in another. The rate at which a solid dissolves in a liquid can be speeded up by heating the liquid, crushing the solid, stirring or shaking the mixture of solid and liquid. Solutions can be concentrated (much solute) or dilute (little solute). The point at which no more solute will dissolve in a solvent is called the saturation point. When materials are "held up" in a fluid the mixture is called a suspension.

Chemists group compounds in various classes according to their chemical properties. Three important classes are acids, bases, and salts. Acids are made up of hydrogen, oxygen and another element, which is usually a nonmetal. Bases are made up of a hydroxide (OH) and another element, which is usually a metal. If an acid and a base combine in the proper amounts, they form a salt and water. The properties of salts vary.

UNIT OBJECTIVES ACCOMPLISHED? FIND OUT.

Part A   Match the numbered phrases 1–10 with the lettered terms.

1. An "atom" that does not have the same number of protons as electrons.
2. The speed in which a solid will dissolve in a liquid.
3. Anything that has weight and takes up space.
4. Equal to the number of protons in each atom of an element.
5. A type of compound often containing hydrogen and oxygen combined with a nonmetal.

a. matter
b. organic
c. element
d. atomic number
e. Periodic Table
f. ion
g. molecule
h. ionic
i. rate of solution
j. acid
k. inorganic

6. Chemical bonding in which electrons are transferred from one atom to another.
7. Matter that contains carbon.
8. A table of all known elements, based on the arrangement of electrons.
9. A simple substance made up entirely of one type of atom.
10. A group of two or more atoms bonded together by the sharing of electrons.

Part B   Choose your answer carefully.

1. The science that studies the makeup and changes in matter is called (a) biology (b) chemistry (c) physics (d) geology.
2. Anything that represents something else is called (a) an atom (b) a molecule (c) a model (d) an ion.
3. The particle that has a negative charge and is found within atoms is the (a) electron (b) proton (c) neutron (d) ion.
4. The lightest of the three elementary particles of an atom is the (a) electron (b) proton (c) neutron (d) ion.
5. An example of an element is (a) water (b) salt (c) oxygen (d) air.
6. If the atom of an element contained 5 electrons, 5 protons and 6 neutrons its *atomic number* would be (a) 5 (b) 6 (c) 11 (d) 16.
7. If the atom of an element contained 15 electrons, 15 protons and 16 neutrons its *atomic weight* would be about (a) 15 (b) 30 (c) 31 (d) 16 a.m.u.
8. Chemists can tell what atoms and how many atoms are in a molecule by looking at (a) the chemical symbol (b) the Periodic Table (c) the chemical formula (d) the a.m.u.
9. All elements whose atoms have the same number of electron shells belong to the same (a) family (b) group (c) period (d) Periodic Table.
10. The chemical name carbon dioxide indicates that the compound contains (a) 2 carbon atoms (b) 1 oxygen atom (c) 3 carbon atoms (d) 2 oxygen atoms.
11. If an atom loses an electron it becomes a (a) positive (b) negative (c) neutral (d) larger   ion.

12. Atoms bond together by sharing or transferring (a) electrons (b) protons (c) neutrons (d) nuclei.
13. A substance in which the atoms of two or more elements are chemically combined is (a) an element (b) a compound (c) a suspension (d) a mixture.
14. Salt water is an example of (a) an element (b) a compound (c) a suspension (d) a mixture.
15. If sugar were dissolved in water the *sugar* would be called the (a) solution (b) suspension (c) solvent (d) solute.
16. A solution can be made more concentrated by (a) adding more solute (b) adding more solvent (c) reducing the amount of solute (d) stirring it.
17. A solute can be removed from a solution by (a) filtering (b) settling (c) stirring the solution (d) distillation.
18. The class of compounds that have a sour taste are called (a) acids (b) bases (c) hydroxides (d) salts.
19. If a substance changes red litmus paper to blue the substance is considered to be (a) an acid (b) a base (c) a salt (d) neutral.
20. An acid plus a base produces (a) hydrogen bubbles (b) a hydroxide and a metal (c) a salt plus water (d) a stronger acid.

Part C Think about and discuss these questions.

1. Compare elements, compounds, and mixtures. Use atomic and molecular models to illustrate your discussion.
2. Compare the two types of atomic bonding that were discussed in this unit. Use atomic models to illustrate your discussion.
3. Why is a solution such as salt water considered to be a *mixture* rather than a *compound?*
4. Compare the properties of acids, bases and salts. What type of elements are found in each?

MATTER AND ITS CONSERVATION

The unit opening photograph shows junk cars traveling up a conveyor into a mill that shreds them into scrap steel. As more and more of the material resources of planet Earth are used up, recycling projects, such as this, are becoming more and more important.

YOUR OBJECTIVE: To realize that physical and chemical properties determine the usefulness of each kind of matter; to investigate physical and chemical change, conservation of mass and simple chemical reactions and equations.

How would you describe a Winesap apple to someone who had never seen one? You might tell the person its color, shape, odor, and taste. You would list the properties of the apple. **The properties of a substance include any quality or characteristic used to describe that substance.**

The four properties — color, shape, odor, taste — are called **physical properties.** Other physical properties include *hardness, density, luster, and the freezing* and *boiling* points of a substance. The freezing point of a substance is the temperature at which it changes from a liquid to a solid. Similarly, the boiling point is the temperature at which a substance changes from a liquid to a gas.

Chemical properties of a substance describe how it interacts with other substances. A chemical property of oxygen is that it will combine with hydrogen to form water. A chemical property of limestone is that it will release hydrogen bubbles when a drop of acid is placed on it.

The physical and chemical properties of a substance determine what it is used for. Steel has the physical property of being very strong. So, steel is used for building bridges. Steel is not used for building airplanes. It is too heavy. Instead, lighter-weight aluminum is used to build airplanes. Aluminum has the physical property of being much less *dense* than steel.

A physical change takes place when any property of a substance changes but no new substance is formed. When a rubber band is stretched, a property — its length — changes. Since no new substance is formed, stretching a rubber band is a physical change.

If butter stands in sunlight in a warm room, it will melt and the solid will become a liquid. Its shape has also changed, but it is still butter. No new substance is formed. The melting of butter is a physical change.

Strike a piece of rock hard enough with a hammer and it will break into smaller pieces. No new substance is formed. The breaking of a rock into smaller pieces is a physical change. Simply changing the *shape, form* or *size* of any substance is considered a *physical change.*

A chemical change occurs when a substance breaks down to form new substances. A chemical change also occurs when two or more substances combine to form a new substance or substances. Fre-

---

Hardness Refers to the resistance of a substance to being scratched or cut.

Density How much of a substance is in a given unit volume.

Luster Refers to the way a substance "shines" when light reflects from it.

quently both breaking down and combining occur in the same reaction. Iron sulfide (a solid made up of iron and sulfur) undergoes a chemical change when mixed with hydrochloric acid (a liquid made up of hydrogen and chlorine). In the process, iron sulfide breaks down releasing iron and sulfur; and, the hydrochloric acid breaks down releasing hydrogen and chlorine (FIG. 1-1a). Then, the sulfur and the hydrogen combine to form hydrogen sulfide (FIG. 1-1b). Hydrogen sulfide is a poisonous gas that smells like rotten eggs. Iron chloride is a liquid. **A chemical change takes place when at least one new substance is formed. In a chemical change, the properties of any** *new* **substance are different from the properties of the substance or substances that produced the new substance.**

Suppose that you put an ice cube in a beaker and place it on the pan of a balance. And, suppose that you cover the beaker to prevent any particles of matter from entering or leaving the beaker. You then balance the mass of the ice cube and covered beaker by placing weights on the other pan (FIG. 1-2a).

If you leave the setup as it is until the ice melts, you will find that the mass of the liquid water is the *same* as that of the frozen water (FIG. 1-2b). Since the *mass* of the water *does not change*, as the ice

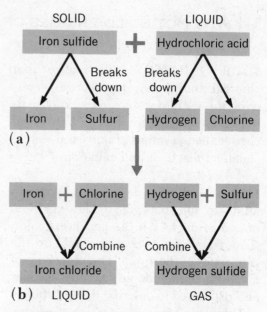

FIGURE 1-1 In this reaction (a) two substances break down to produce four elements (b) which in turn combine to form two new substances.

changes to liquid, you can say that the mass is *conserved*.

The melting of an ice cube is an example of a *physical* change. **Scientists have shown that during any physical change mass is conserved.**

In an ordinary chemical change mass is also conserved. Suppose you place a test tube containing a silver nitrate solution upright inside a stoppered flask that

---

Reaction Combining or breaking up of a substance or substances that results in a new substance(s) being formed.

Mass The amount of matter in an object.
Ordinary chemical change
Change that does not involve change in the nucleus of an atom.

Conserved Saved.

FIGURE 1-2 The mass of a given amount of water does not change when the water changes state.

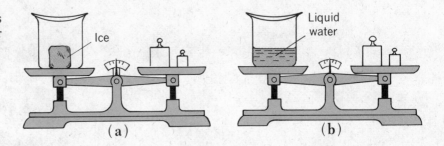

(a) (b)

contains a solution of sodium chloride. And, you then put the flask and its contents on one pan of a scale and balance it by placing weights on the other pan (Fɪɢ. 1-3a).

You next "spill" the silver nitrate from the test tube into the sodium chloride solution. A chemical change takes place. You will observe that *a new substance forms* and settles to the bottom of the flask. This new substance is silver chloride. You will observe also that the setup is still balanced (Fɪɢ. 1-3b). The total mass inside the flask does not change. In an ordinary chemical change, mass is conserved.

When a chemical change occurs the process that causes it is called a chemical reaction. **Every chemical reaction involves energy change.** In some chemical reactions energy is *given off*. This type of reaction is called an *exothermic* reaction. The burning of wood is an example of a chemical reaction that is exothermic. Heat and light energy are given off by the reaction.

The sudden release of energy in an explosion of dynamite is also an exothermic chemical reaction. Most exothermic reactions do not give off energy as quickly as in an explosion. The energy in most exothermic reactions is released so slowly that it is hardly noticeable. The rusting of iron is an exothermic reaction. This chemical reaction occurs so slowly that it is hard to detect any release of energy.

Energy The ability to do work; that is, to exert force through a distance.

FIGURE 1-3 There is no change in mass in this chemical change. Why is a stopper placed on each flask? How do you know that a chemical change took place?

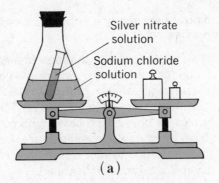

Silver nitrate solution

Sodium chloride solution

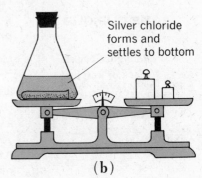

Silver chloride forms and settles to bottom

(a) (b)

FIGURE 1-4 This is a photograph taken of the *USS Albany* as three missiles were fired in salvo (at one time) from the ship. From what you can observe were the three reactions endothermic or exothermic? *(U.S. Navy)*

Some chemical reactions will not take place unless energy is *added*. Such chemical reactions are called *endothermic* reactions.

Water can be broken down into oxygen and hydrogen by *adding* electrical energy. Mercuric oxide can separate into oxygen and mercury if *heat energy* is added to the oxide (FIG. 1-5). Both reactions are examples of endothermic reactions.

Chemical reactions can be described by the use of word equations. The word equation shown is read: sulfur trioxide *plus* water vapor produces (yields) sulfuric acid.

sulfur trioxide + water vapor → sulfuric acid

Sulfur trioxide reacts with water vapor to produce sulfuric acid. The sulfur trioxide and the water vapor are called the *reactants*. Sulfuric acid is called the *product*.

> *Equation* In chemistry, a symbolic representation of a chemical reaction.

BE CURIOUS 1-1: **Find out how readily a substance can be identified by its properties.**

Reference books
Clock or watch

Make a list of properties that describe each of these substances: gold, iron, gasoline, butter, water, diamond, wood, glass, oxygen, and a substance of your own choice. List as many properties as you can think of for each substance. Tell whether each property is a *chemical* or *physical* property. Find out how well the properties you have listed describe a substance. Read your list of properties for one substance to your classmates. Note the time. Discover how readily the substance is identified.

Can you identify common physical and chemical changes?

Reference books

Use any available source (observation, reference books) to tell what changes occur in each of the events in the list that follows. Identify each as a *physical* change or a *chemical* change. Explain your answer.

melting of an ice cream cone
rusting of a nail
window glass being broken
digestion of food

wood burning in a fireplace
bending a wire
gasoline exploding
rising of bread dough

This reaction can also be described by a *chemical equation*. The *chemical* equation tells us more about the reaction than does the *word equation*.

$$SO_3 + H_2O \longrightarrow H_2SO_4$$

FIGURE 1-5 In this reaction mercuric oxide is the reactant. What are the products?

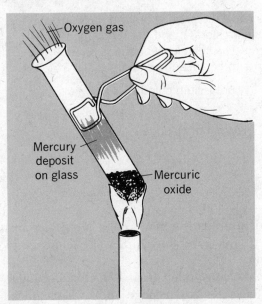

Oxygen gas

Mercury deposit on glass

Mercuric oxide

A chemical equation shows the numbers and kinds of atoms that take part in a reaction. It also shows how the various atoms are arranged before and after a chemical reaction. The *total* of matter and energy on one "side" of a reaction is equal to the *total* of matter and energy on the other "side". Combinations of symbols of elements known as *formulas* show what a substance is made of. The chemical formula SO_3 stands for sulfur trioxide. It tells us that sulfur trioxide contains *one* atom of sulfur (S) combined with every *three* of oxygen (O). The chemical formula H_2O shows that *two* atoms of hydrogen (H) combine with every *one* of oxygen to form water.

Suppose that you wished to write a word equation showing that hydrogen and oxygen react to form water. It could be written this way:

hydrogen + oxygen $\longrightarrow$ water

Atoms Small particles referred to as the "building blocks" of matter.

Hydrogen atoms do not exist naturally as single atoms. They occur in groups of *two* (H_2) — a *diatomic molecule*. Oxygen also is found as a diatomic molecule (O_2).

❋ The chemical equation for the reaction of hydrogen and oxygen to form water indicates that *two* atoms of hydrogen and *two* atoms of oxygen react to form water.

$$H_2 + O_2 \longrightarrow H_2O$$

Note that the molecule of water contains only *one* atom of oxygen. Since there are *two* atoms of oxygen on the left side of the equation and only *one* on the right, this equation is *unbalanced*.

❋ You can balance an equation by the use of *coefficients*. You can place the coefficient 2 before the formulas for hydrogen and water and balance the equation.

$$2H_2 + O_2 \longrightarrow 2H_2O$$

❋ The coefficient *2* tells us that there are *two molecules* of H_2 in the equation — a total of *four* hydrogen atoms. There is *no* coefficient placed before the molecule of O_2. This indicates that there is only *one* O_2 *molecule* — a total of *two* oxygen *atoms*.

❋ This balanced equation shows that *two* molecules of water are formed. There are a total of *four* hydrogen atoms and *two* oxygen atoms in the two molecules. The equation is balanced because there are the same number of each type of atom on both sides of the equation.

> *Coefficient*   In chemistry, a number placed before a substance to indicate the number of molecules of the substance.

❋ BE CURIOUS 1-3:   **Find out if you can balance an equation.**

The following equations show the production of potassium chloride and oxygen from potassium chlorate; and the production of salt from sodium and chlorine. The chemical equations are not balanced. Use coefficients to balance the equations.

potassium chlorate $\longrightarrow$ potassium chloride + oxygen
$KClO_3 \longrightarrow KCl + O_2$

Sodium + chlorine $\longrightarrow$ sodium chloride (salt)
$Na + Cl_2 \longrightarrow NaCl$

OBJECTIVE 1
ACCOMPLISHED?
FIND OUT.

1. How does a chemical property differ from a physical property of a substance? Give several examples of chemical and of physical properties.

2. How does a physical change differ from a chemical change? Give an example of a physical change and of a chemical change.

3. How do the properties of a substance produced by a chemical change compare with those of the substances that produced it?

4. What is meant by conservation of mass during a physical or chemical change?

5. How does an exothermic chemical reaction differ from an endothermic chemical reaction?

6. Study the chemical equation $Zn + H_2SO_4 \longrightarrow ZnSO_4 + H_2$. Answer the following questions. (a) Which substances are the reactants and which the products? (b) What does the arrow in the equation mean? (c) How many hydrogen atoms are shown on the left side of the equation? on the right side? (d) How many sulfur atoms are shown on the left side of the equation? on the right side? ✸ (e) Is the equation balanced?

✸ 7. Use coefficients to balance the following two chemical equations: $PbS + O_2 \longrightarrow PbSO_4$; $N + H_2 \longrightarrow NH_3$

2 | FUELS AND THE ENERGY CRISIS

YOUR OBJECTIVE: To understand what is meant by the terms renewable and nonrenewable resources, fuel, and combustion; to identify the properties of various fuels; to evaluate fuel needs and short- and long-term solutions to the fuel-energy crisis.

An industrialized nation, such as the United States, needs a very, very large amount of fuel to maintain its life style. **Fuels provide energy. And, energy provides the ability to do work.** However, a serious problem exists and will exist for some time — certainly until the beginning of the 21st century. The problem is that

| *Work* To a scientist, work is done when an external force acts on an object through a distance. |

there is an increasing demand for energy and a decreasing supply of fuels.

In an industrialized nation, fuel is the most important resource (not considering elements necessary to maintain life such as oxygen, hydrogen). Coal, natural gas, petroleum, and wood are fuels. **Fuels are substances that produce large amounts of heat when burned.** Fuels are natural resources. Of the fuels commonly in use, wood is the only one that is *renewable*.

Many of the materials you use come from renewable resources. **A renewable resource is one that can be replaced naturally in a reasonable amount of time.** Consider lumber. Suppose that a tree is cut down and made into lumber. Another tree can grow to replace the cut tree. A tree is a renewable resource. Lumber is made from trees. Lumber is a renewable material. But, that does not mean that

FIGURE 2-1 A research forester shows a cut 460-year-old tree (left) that grew naturally; and a 60-year-old cut tree that was grown in a controlled environment. Compare the trees. Do you think scientific research "pays off"? Explain. (*June Malcolm for The New York Times*)

trees can be cut down without limit to provide lumber.

It takes time for a tree to grow. When lumberjacks cut an area of forest, it takes from 30 to 200 years for the area to *regenerate* naturally (FIG. 2-1). The length of time it takes a forest to regenerate is determined by the kinds of trees native to the forest and the nature of the environment.

People in this country and all over the world must plan carefully in order to *conserve* renewable resources. **The use of renewable resources must be planned so** that an adequate supply will be continuously available. Nations must be even more careful to conserve *nonrenewable* resources. **Nonrenewable resources are those that cannot be replaced naturally in a reasonable amount of time.**

Petroleum is a nonrenewable resource. It takes *millions* of years (and a particular set of conditions) for petroleum to form within the earth. But, without careful planning it is quite possible to exhaust petroleum resources in a few *hundred* years!

The elements and compounds taken from the earth's crust are usually nonrenewable. Nonrenewable resources include materials essential to our way of life — resources such as iron, aluminum, petroleum and coal. Although iron and aluminum are nonrenewable materials, they *can be recycled*. They can be used over again and again by melting scraps and forming new metal. Petroleum and coal *cannot* be recycled. There is generally no way to use them more than once as fuel.

When coal — or any carbon fuel — burns, the carbon in the fuel reacts with oxygen from the air. This reaction forms gases as carbon and oxygen atoms combine. The gases are *carbon dioxide* (CO_2) and *carbon monoxide* (CO).

The burning, or combustion, of coal is an example of a very important kind of chemical reaction, an *oxidation reaction*.

Regenerate To reproduce; grow again.
Conserve In reference to materials or resources, to keep from wasting or exhausting.

Carbon fuels Some common fuels that contain the element carbon are coal, coke, charcoal, oil, gasoline, kerosene, propane gas, alcohol.

Oxidation reaction A reaction in which there is a loss of electrons by an atom. Oxygen does *not* have to be involved in an oxidation reaction, but often is.

In this reaction, oxygen combines with carbon to form an oxide. It is an *exothermic* reaction. The reaction takes place quickly and gives off heat. This type of oxidation is sometimes called a *rapid oxidation* reaction. The amount of heat given off is very noticeable. Combustion usually also releases light energy along with heat energy.

Most combustion occurs when *oxygen* from the air combines with the *carbon* or *hydrogen* in a fuel. **Combustion will begin only when the fuel has reached a certain temperature.** This temperature is called the *kindling temperature* of the fuel. The combustion stops when the fuel is cooled below the kindling temperature.

Different kinds of fuels have different kindling temperatures. As you know gasoline, kerosene, or the gas you cook with catch fire when they are heated by a match flame. These fuels have low kindling temperatures. Logs in a fireplace usually do not catch fire when you put a match to them. To get logs to burn you usually start a small fire beneath the logs to heat the logs to kindling temperature. A small pile of wood shavings or twigs or pieces of newspaper can be ignited with a match flame and can be used to raise the logs to kindling temperature. Can you give two reasons why small, thin pieces of wood or of newspaper will ignite with a match and a log will not?

Water is a common product of combustion. It is formed only when certain fuels are burned. The oxidation of the hydrogen of a fuel produces water (FIG. 2-2).

> *Product* In chemistry, anything formed as a result of a reaction.

If there is *much oxygen* present when a fuel burns, *complete combustion* may take place. Carbon dioxide is a product of complete combustion. These equations show the complete combustion of methane (natural gas). The arrow next to CO_2 in the equation indicates that the substance is a gas.

methane + oxygen ⟶ carbon dioxide
+ water

$$CH_4 + 2O_2 \longrightarrow CO_2\uparrow + 2H_2O$$

If there is a *poor supply of oxygen* present when a fuel burns, *incomplete combustion* may take place. If so, carbon *monoxide* — not carbon *dioxide* — forms. These equations show the *incomplete* combustion of methane gas.

methane + oxygen ⟶ carbon monoxide
+ water

$$CH_4 + O_2 \longrightarrow CO\uparrow + H_2O$$

FIGURE 2-2 This fuel, the candle wax, contains hydrogen. When the candle burns, water (H_2O) forms. Where does the oxygen come from to form the water molecule?

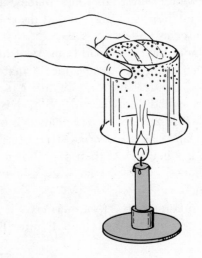

Incomplete combustion can also produce soot. *Soot* is unburned carbon that is released from a fuel. Soot is black. It is often responsible for the black you see in the smoke from some chimneys and in automobile exhaust. *Ash* is the remains of the noncarbon material (except H) that was present in the solid fuel before it was burned.

Scientists think that coal formed from the remains of plants that grew millions of years ago. Through millions of years, heat and pressure from surrounding earth materials slowly changed buried layers of plant remains into coal. In the process of becoming coal, plant remains go through several different changes. The first change forms *peat*. Peat is a brown spongy substance. Peat does not make a very good fuel. It produces large amounts of smoke when it burns. Peat does not produce much heat unless large amounts of it are burned at one time. Much ash remains after peat has been burned. If you have a garden you may be familiar with one form of peat. That is peat moss. Gardeners mix peat moss with soil to help the soil retain water. In some places, as in Ireland, blocks of pressed and dried peat are used as fuels.

Under the right conditions, peat found in the earth can change into a material called *lignite* (ˈlig·nīt). Lignite is sometimes called "brown coal." But, it is *not* coal. It is not spongy like peat. It is fairly hard and has a *woodlike* structure. Lignite is a better fuel than peat, but it still produces much smoke and ash when burned.

Bituminous (bə·ˈt(y)ü·mə·nəs), or *soft coal* is a better fuel than lignite. Bituminous coal is the most abundant type of coal. It is black in color. Many smelly gases are produced when it is burned. Bituminous is considered a *low-grade coal*. Low grade coals have a lot of moisture in them and little carbon. Therefore, they have low heating value.

Anthracite (ˈan(t)·thrə·ˌsīt), or hard coal is formed from bituminous coal. Over hundreds of years, heat and pressure within the earth change soft coal into hard coal. Hard coal has a higher percentage of carbon in it than does soft coal. Hard coal also contains less water. It is *high-grade coal*. Anthracite coal produces less smoke and ash. It also burns more slowly than does soft coal.

High-grade coal produces more heat energy per ton than does low-grade coal. However, supplies of efficient high-grade coal are not sufficient to meet our increasing energy demands. So, low-grade coal will be used more and more to supply energy needs.

A flame is produced by a burning gas. Yet, a flame can be seen when *solid* fuels such as coal or wood are burned. The flame around the coal or wood is produced by the gases driven from the hot solid fuels. Suppose that all gases could be driven from coal or wood. The remaining solid fuel would then burn *without a flame*.

Most gases can be driven from coal when the coal is heated in a test tube or container in which there is no oxygen (FIG. 2-3). Because oxygen is *not* present,

Flame The glowing gaseous part of a fire.

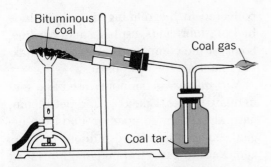

Bituminous coal

Coal gas

Coal tar

FIGURE 2-3 When bituminous coal is heated in a *Pyrex* test tube over a hot flame, a flammable gas escapes. As the gas cools in the collecting bottle, some condenses and forms a tarlike liquid. If heating continues until all gases are driven from the coal, the coal is changed to coke.

the hot coal will *not* burn. But, gases will be driven from the coal as it is heated. Some of the gases change into a liquid as they cool in the collecting bottle. Other gases escape through an outlet tube. These burn with a flame when ignited.

The solid material that remains in the test tube is called *coke*. Coke can be burned. When it burns, there is almost no flame. Why is there little flame when coke burns?

Coke has a higher carbon content than coal. It produces a higher temperature when it burns. Coke is used as a source of carbon for making steel. The gases

BE CURIOUS 2-1:   **Find out why combustion stops.**

Part A

You know that in order for combustion to take place you must (a) have a fuel (burnable material); (b) raise the fuel to kindling temperature (supply heat); and (c) have oxygen present. Explain why each of the following can be an effective way of stopping combustion.

1. Throwing a rug or blanket over burning waste in a wastebasket
2. Spraying water on a grass fire
3. Spraying water on unburned grass that is in the path of a burning grass fire
4. "Blowing out" a candle
5. Turning off a burner on a gas stove
6. Throwing sand on a burning campfire

Part B

Reference books

Use outside sources to find out how each of the following types of fire should be stopped and what precautions should be taken while stopping the fire.

1. Electrical fires
2. Burning gasoline
3. Burning clothes on a person
4. Grease burning in a frying pan

driven from coal while making coke also have many uses. The gases are used to produce coal tar, ammonia, coal gas and various oils.

Charcoal is produced in a similar way as is coke. Charcoal is made by heating wood in an area free of oxygen. Gases are driven from the wood. The solid remains, charcoal, can be used as a fuel. Charcoal burns at a higher temperature than wood. When charcoal burns, do you think it will burn with less, more, or about the same amount of flame as a similar–sized piece of wood?

Scientists think that petroleum, also called crude oil, probably formed from marine life that died millions of years ago. Through many, many hundreds of years as the marine life was buried by many other layers of rock, heat and pressure changed the remains to the liquid known as petroleum. The petroleum became concentrated in certain areas underground. It gathered in *porous* rocks and was held there by *nonporous* rocks that surrounded the porous rocks.

Petroleum is a mixture of many different compounds. The process used to separate these compounds is called *distillation* or *refining*. It is a heating process. In refining, the temperature of the petroleum is increased in *steps*. At each of these steps a different compound of petroleum is changed to a gas and is driven off and collected.

Each gas changes back into a liquid when it cools. Some of the compounds collected in the refining process include fuel oil, lubricants, asphalt, and gasoline. Some of the compounds are used in making materials such as plastics.

Natural gas is commonly used as a fuel in many homes. Like coal and petroleum, natural gas formed underground from organic matter that lived millions of years ago. Natural gas is trapped in porous rock formations beneath the earth's surface. Natural gas is generally found in areas where petroleum is found. The gas can be released by drilling wells down into the rock in which it is trapped.

Natural gas is a mixture of different compounds of carbon and hydrogen. Most natural gas is methane (CH_4). Many other compounds of carbon and hydrogen in the form of gas are used as fuels. Two often used are the "bottled" gases *propane* (C_3H_8) and *butane* (C_4H_{10}).

The fuel of the future could very well be hydrogen. Many scientists think that hydrogen will replace many fuels in use today. At present, hydrogen is used as a fuel in a few instances. One use is as a fuel in rockets. Perhaps some day your home may be heated and your automobile may be powered by hydrogen.

Hydrogen is a *very abundant* material. It is one of the elements that make up water (H_2O). By passing an electric current through water, hydrogen can be easily separated from oxygen and collected. Hydrogen can be transported and stored in much the same way as natural gas. It can be transported in underground pipes and stored in large tanks. Both are *inexpensive* ways of handling a fuel.

Using hydrogen as a fuel has advantages. It is not necessary to provide ventilation when hydrogen is burned. If hydro-

Marine Of or relating to the sea.
Porous Containing tiny spaces.

Find out what products of coal and petroleum you use.

Use any sources available to make a complete listing of products made from coal and petroleum. Use the lists to help answer the following questions. Which, if any, of the items on your list do you need to stay alive and well? Which, if any, of the items on your list do you consider "luxury" items? Which items would you be willing to do without? What substitutes could you find for any products on your list? How would your life style change if coal and petroleum were no longer available? Discuss these questions with your classmates.

gen were burned in a home furnace, a chimney would not be needed. There are *no pollutants* formed when hydrogen burns. Water vapor is the only product of combustion. Water vapor can be released into the environment and no harm is done. In fact, the water vapor from the combustion of hydrogen fuel could be used to humidify the air in the home.

The use of hydrogen as a fuel does have some drawbacks. Hydrogen is more dangerous to work with than fuels now commonly used. It is more explosive. Extra safety features must be built into the equipment transporting, storing or using hydrogen. Since hydrogen has no odor, a leak of the gas might not be easily detected. But, some substances could be mixed with the hydrogen to provide an odor. Hydrogen also burns with a colorless flame. Thus, the flame could not be seen. But again, some substance could be mixed with the hydrogen to color the flame.

Hydrogen may indeed become "the fuel of the future." But, that "future" will be quite a time arriving. The "hydrogen fuel program" is still in the *research* stage. Many problems remain to be solved to assure efficient production, storage, trans-

portation, and use of this high-energy fuel. When research is completed and a pilot program is in action, it will take at least another five to seven years to move into commercial production.

An immediate answer to the fuel-energy crisis seems to be to return to coal. There are still large reserves of coal in this country. Of course, much of the coal is low grade and so when it is burned it produces pollutants. But, research is turning up ways to use coal that will be less harmful to the environment.

The process that appears to offer the best solution in the immediate future is to obtain a gas (fuel) from coal by the *coal gasification process*. One such process is Project *Bi-Gas*. A pilot plant for the project is under construction. The process should be in commercial production by 1980 or 1981.

> **Research** Systematic investigation to find out something. In science frequently an experiment is involved. An experiment is a test or trial that involves a control (check), which provides a standard for comparison.

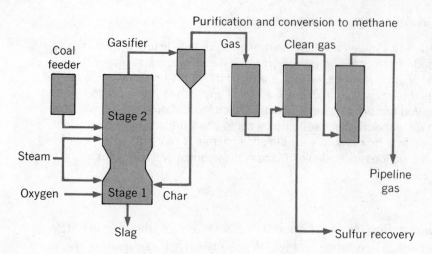

Coal feeder

Gasifier

Purification and conversion to methane

Gas

Clean gas

Stage 2

Steam

Oxygen

Stage 1

Char

Slag

Pipeline gas

Sulfur recovery

FIGURE 2-4 A simplified diagram of the *Bi-Gas* gasification process. (*American Gas Association*)

✳ Project *Bi-Gas*, developed jointly by the Department of the Interior and the American Gas Association, obtains a gas from coal that is rich in methane. It is very similar to natural gas and so is suitable for home heating and industrial use. Study the two-stage process (FIG. 2-4). In the first step, char obtained by heating coal passes upwardly while reacting with steam and oxygen. In the second stage, the gas so produced reacts with coal and steam. Purification and conversion to *methane* follow. The methane gas may be used as a fuel or changed into a clean, liquid boiler fuel or to methyl alcohol.

✳ You may wonder what is so new about this process. It seems very much like the process by which coal is heated to produce coke (FIG. 2-3). But, there are important differences. Relatively low temperatures are used to produce coke. High temperatures of from 1700°–1900°F (940°–1100°C) are used in the *Bi-Gas* process. Project *Bi-Gas* is also carried out under high pressure, and the coal is heated not in air but in a gaseous mixture that has a high hydrogen content. The

result is that the *Bi-Gas* process produces a high-methane gas *low* in carbon monoxide and *high* in carbon dioxide. The gas produced in the "coke process" is just the opposite. It is *high* in carbon monoxide and *low* in carbon dioxide. The "coke" gas is therefore toxic and low in heating value — not suitable for home and industrial use. The *Bi-Gas* high-methane gas is nontoxic, so can be used in the home; and, is high in heating value so it is suitable for both home and industrial fuel use.

The United States is not the only country that is seeking new energy sources. The population of the world will double by the year 2000. Every year, machines are being made available for use in parts of the world where they have previously not been used. The energy needs of this large, mechanized world population of the 21st century will be five times as great as they are now. Fuels will

Pressure The force acting upon a surface per unit of area.

supply much of this energy. The question at hand is not whether energy production for the world should be increased. The problem is how to increase it with the least number of harmful side effects. In this short section, it has been possible only to touch upon some of the problems involved and upon some possible solutions. You can further inform yourself and keep up-to-date with the fuel-energy problem by reading newspapers and magazines. You will find many articles.

OBJECTIVE 2 ACCOMPLISHED? FIND OUT.

1. What is a fuel?
2. What is meant by combustion?
3. How does complete combustion differ from incomplete combustion?
4. How do scientists think coal and petroleum formed?
5. How are coke and charcoal produced?
6. Give the names of some gases that are used as fuels.
7. Why may hydrogen be "the fuel of the future"?
* 8. Explain the advantages of Project Bi-Gas.
* 9. What problems can you foresee as the result of switching to coal as a major fuel source?

3 | METALS AND ALLOYS

YOUR OBJECTIVE: To appreciate the usefulness of metals and to describe the properties of metals that explain why they are useful; to find out what is meant by the terms corrosion, alloy, pure metal and physical metallurgy.

About two-fifths of all the known elements are metallic. Look at the squares shown in color in the Periodic Table of Elements (FIG. 3-1). How many elements can you name by referring to the symbols?

Such metals as aluminum, iron and sodium are very abundant in the earth's crust. Other metals such as nickel, copper and zinc are not so abundant. There is over one thousand times as much aluminum in the earth's crust as there is nickel, copper or zinc.

Some metals are very rare in the earth's crust. Gold, silver and platinum are among the rare metals. Aluminum is many millions of times more abundant than any of these three.

Elements and mixtures found in the earth's crust are essential to maintaining a highly-developed, industrialized society such as ours. You have investigated one aspect of this resource — fuels. Next to fuels, metals are perhaps the most valuable group of materials obtained from the earth's crust. **Metals are valuable materials because they have unique physical and chemical properties that make them suitable for a wide variety of uses.**

IA																	0
H	IIA											IIIA	IVA	VA	VIA	VIIA	He
Li	Be											B	C	N	O	F	Ne
Na	Mg	IIIB	IVB	VB	VIB	VIIB		VIIIB		IB	IIB	Al	Si	P	S	Cl	Ar
K	Ca	Sc	Ti	V	Cr	Mn	Fe	Co	Ni	Cu	Zn	Ga	Ge	As	Se	Br	Kr
Rb	Sr	Y	Zr	Nb	Mo	Tc	Ru	Rh	Pd	Ag	Cd	In	Sn	Sb	Te	I	Xe
Cs	Ba	Lantha-nide series	Hf	Ta	W	Re	Os	Ir	Pt	Au	Hg	Tl	Pb	Bi	Po	At	Rn

FIGURE 3-1   Periodic Table of the Elements: The elements on the left of the table are strongly metallic. The metallic properties gradually lessen from element to element as you go from left to right across the table. (Note: nonmetallic elements 87–106 are not shown.)

Useful *physical properties* of metals are that they are *hard, malleable, ductile;* have *high melting points,* and *conduct heat* and *electricity* well.

Hardness refers to the ability of a substance to withstand scratching. Most metals are hard to scratch. If a substance is hard to scratch, it *usually* means the substance will also be hard to break. It will be strong. Most metals are strong. Many metals have a high tensile strength. **Tensile strength refers to the degree of resistance to the force needed to tear a strand of material while pulling it apart.** An iron cable has a higher tensile strength than a cotton rope.

A material is said to be malleable if it can be rolled or hammered into thin sheets without the material cracking. Most metals are malleable. Gold is very malleable (Fɪɢ. 3-2). Thin sheets of gold are used to make the "gold leaf" lettering seen on many office windows. Tin and aluminum can easily be made into thin foils.

Most metals are ductile. **A material is said to be ductile if it can be drawn into long thin wires without breaking.** Iron can be drawn into thin wire. The wire can be used in making such products as fencing material or nails. Aluminum and copper are often drawn into wire, which is used to carry electricity.

Most metals melt at a very high temperature. All but one are solids at room temperature. Mercury is a silver-gray liquid at room temperature.

Most metals are good conductors of heat and electricity. This means that heat and electricity can pass through them easily. Copper and iron are used to make cooking utensils because they conduct heat well. Copper and aluminum are used to make wires for electrical systems because they conduct electricity well.

Because metals are opaque and reflect light well they have a high luster. Although metals have a high luster, they can

Luster   The appearance of an object determined by the light it reflects.

tarnish (lose luster) when continually exposed to air. Tarnishing of a metal is called *corrosion*. **Corrosion is caused by the reaction of a metal with some gas in the air.**

One of the most common examples of corrosion is the rusting of iron. Iron and oxygen react to form a reddish-brown flaky substance called *rust*. Rusting is a continuing problem with many objects made from iron. Unless stopped, rusting can destroy a piece of iron.

Aluminum can also react with oxygen from the air to produce a white *aluminum oxide* (Al_2O_3). The corrosion of aluminum does not consume — "eat up" — aluminum as rusting does iron. Aluminum oxide forms a protective coating on the surface of the aluminum that prevents further oxidation.

The white coating of aluminum oxide that forms on aluminum acts as a barrier to further corrosion. The coating of rust that forms on iron does not protect the iron under it. Because iron rust is a soft, light substance, it quickly flakes off exposing the iron beneath it to the air. The newly exposed iron then begins to rust. Because rust flakes a rusting-flaking-rusting cycle can go on until the iron is destroyed. For this reason, corrosion is usually much more harmful to iron than it is to most of the common metals.

When a natural coating does not protect a metal from corrosion, it is possible to apply a coating. Sometimes paint is used as a coating to prevent corrosion. At other times, a thin coating of grease or oil is used to prevent corrosion.

Sometimes one metal is protected from corrosion by coating it with a thin layer of another metal. To prevent iron from

FIGURE 3-2 This gilded statue of Prometheus overlooks the main plaza of Rockefeller Center in New York City. Periodically when weathering and corrosion dull the gold leaf covering, the gold leaf is replaced. Observe how thin and pliable the gold leaf is. A skilled technician is carefully peeling leaf from the toe of the statue. The plastic heated enclosure was built because the new gold leaf can be applied only at a warm temperature. (*Neal Boenzi for the New York Times*)

Compare the rusting of steel wool under various conditions.

Part A

Steel wool
6 beakers
6 test tubes
Water
Petroleum jelly or other
   protective
Plastic wrap or other
   wrap

Place a small amount of dry steel wool in the bottom of a test tube. Place the test tube upside down in a beaker of water (see the figure at the left). Moisten a small amount of steel wool with water and place the wet steel wool in another test tube. Also place this test tube upside down in a beaker of water.

Decide on a way to try to keep two other pieces of steel wool from rusting. Perhaps you could rub petroleum jelly on some of the steel wool or wrap some in plastic food wrap or other covering. Place these pieces in test tubes and invert in a beaker of water.

Do not disturb the setups for several days. Then observe the contents of each test tube. What do you observe in each test tube? Why does the water level rise in the test tubes? In which test tube did it rise the most? the least? Which steel wool shows the most sign of rust? the least sign?

Part B

Place a small amount of dry steel wool in the bottom of one test tube, and a small amount of water-wet steel wool in another test tube. Place *each* test tube upside down in a *dry*, empty beaker. Do not disturb the setups for several days. Then observe the contents of each test tube. Has the steel wool rusted in either test tube? Explain what you observe.

rusting, it is often coated with a thin layer of *zinc*. This process — coating with zinc — is called *galvanizing*. Iron coated with zinc is called *galvanized iron*. Galvanized iron is often used on rain gutters, iron pipes and garbage cans. Why do those objects need a protective coating?

A few metals can be found in the earth's crust in their pure form. **A pure metal is made up of atoms of only one element.** Pure metals are called *native* metals. Gold and silver are found as native metals. Copper can be found as a native metal or combined with other elements.

Often metals are not used in their pure form. Instead, two or more metals may be mixed, or alloyed together. They are usually mixed together while they are hot and in their liquid states. When they cool, they form a new material. The new material is called an *alloy*.

Metals are alloyed together to form special materials with certain desired properties. Suppose that you wanted to make inexpensive and strong knives, forks and spoons. Iron might be ideal except that it rusts. By mixing small amounts of chromium and manganese with the iron, an alloy can be formed. This alloy is called *stainless steel*. Stainless steel is strong, cheaper than silver and does not tarnish. It is ideal to use to make inexpensive knives, forks and spoons.

Among the properties that can be developed by alloying metals are strength, corrosion resistance, resistance to wear due to friction, and hardness. The science that is concerned chiefly with the uses of metals and the alloying of metals is called physical *metallurgy* (ˈmet·əl·ˌər·jē).

Brass and *bronze* are both alloys of copper. **Brass is a mixture of copper and zinc. Bronze is made by alloying copper with tin.** Brass and bronze are harder than any of the metals mixed with them. These alloys are often used in making plumbing fixtures and decorative items.

Aluminum has the property of being very light. It might seem to be a good metal for building aircraft. However, pure aluminum does not have the strength needed in aircraft structures. Aluminum is alloyed with other metals to make it strong enough to use in aircraft. **A mixture of aluminum, nickel, cobalt, and iron produces an alloy called *alnico*.** Alnico is an ideal material for making strong and long-lasting magnets such as are used in loudspeakers, telephones and hearing aids.

Solder (ˈsäd·ər) is a very common alloy. **It is a mixture of lead and tin.** It has the property of melting at a fairly low temperature. Solder can be melted with an electric soldering iron. The liquid solder is used to join metals together as it hardens. Your radio or T.V. set has many points at which the parts are held together by a drop of solder.

OBJECTIVE 3 ACCOMPLISHED? FIND OUT.

1. Why are metals perhaps the most valuable group of materials obtained from the earth's crust?
2. Describe each of the following physical properties as they apply to metals: hardness, malleability, ductility, tensile strength.
3. What is meant by the corrosion of a metal? List several ways that the corrosion of metals can be prevented.
4. Explain why rusting is more harmful than many other forms of corrosion.
5. What is a pure metal? Name one metal usually found in the pure state in the earth's crust.
6. What is an alloy? Give some reasons why alloys are developed.
✢ 7. What metals are mixed together in each of the following alloys: stainless steel, alnico, brass, bronze, and solder.
✢ 8. What is the basic difference in the makeup of a pure metal and of an alloy?
9. What is the name of the science that is concerned chiefly with the uses of metals and the alloying of metals?

4 | MORE ABOUT METALS AND NONMETALS

YOUR OBJECTIVE: To find out what an ore is and how some metals are extracted from ores; to identify the properties and uses of some of the most abundant and most important metals and nonmetals found in the earth's crust; to consider the need for conservation of the resources of the earth's crust.

Most metals that occur naturally in the earth's crust are not found in the pure state — as elements. **Most metals in the earth's crust are found in the form of compounds.** Iron and aluminum are often found combined with oxygen as iron oxide (Fe_2O_3) and aluminum oxide (Al_2O_3). Zinc and lead are often found combined with sulfur as zinc sulfide (ZnS) and lead sulfide (PbS). **Such compounds of metals that occur naturally in the earth's crust are called minerals. Minerals are solids, found within or on the earth, that are made up of one or more elements. Minerals can be metallic or nonmetallic.**

Rocks of the earth are made up of mixtures of minerals. When a concentration of one or more minerals is found in rocks, it is referred to as a *deposit*. A deposit of rocks or minerals from which a metal can be removed profitably is called an *ore*. The science that studies ways to remove a metal from its ore is called *process metallurgy*.

A high-grade ore is one from which large amounts of metal can be removed simply and economically. As high-grade ores are used up, the mining industry must turn to less profitable low-grade ores. It costs a lot to get a little metal from a low-grade ore. Usually this is because in a similar amount there is less metal in a low-grade ore than there is in a high-grade ore. Therefore you must obtain and process a *lot* of low-grade ore to get a *little* metal.

The ores of many metals are oxide or sulfide compounds. The way a metal is removed or extracted from an oxide ore differs from the way a metal is extracted from a sulfide ore.

A metal can be removed from an oxide ore by a process called reduction. Reduction can take place when an ore is heated in the presence of a *reducing agent*. The reducing agent reacts with the oxygen of the ore to form a gas. The gas is given off leaving the metal behind. Carbon or carbon monoxide are often used as reducing agents because they react readily with the oxygen of the heated ore to form gases.

The equations that follow show how an oxide of copper might be reduced. The oxide of copper is heated with charcoal. Besides providing heat, the charcoal also provides the reducing agent, carbon. The

---

Process A method of operations in the production of something.

Reduction In chemistry, a reaction in which an atom gains electrons.

Reducing agent The atom or group of atoms that supplies electrons during an oxidation-reduction reaction.

---

carbon reacts with the oxygen from the heated copper ore. The carbon and oxygen combine to form carbon dioxide gas (gas indicated by upward pointing arrow). Pure copper remains behind.

Copper oxide + carbon (from charcoal)
$$\longrightarrow \text{copper} + \text{carbon dioxide}$$

$$2\,CuO + C \longrightarrow Cu + CO_2\!\uparrow$$

Metals are commonly removed from sulfide ores by a process called roasting. In the roasting process the sulfide is heated in the presence of air. The oxygen from the air replaces the sulfur in the ore. The oxygen can then be removed from the ore by reduction.

The following equations show how a sulfide of copper might be roasted. The sulfide is heated in the presence of air. Some oxygen of the air replaces the sulfur in the ore. An oxide of copper is produced. Some of the remaining oxygen combines with the sulfur released from the heated ore to form sulfur dioxide gas.

Copper sulfide + oxygen (from air)
$$\longrightarrow \text{copper oxide} + \text{sulfur dioxide}$$

$$CuS + O_2 \longrightarrow CuO + SO_2\!\uparrow$$

The oxygen can then be removed from the copper oxide by the process of reduction. Pure copper is left behind.

Aluminum is the most abundant metal found in the earth's crust. It is found in many rocks and clays on the earth. The commercially important ore of aluminum is called bauxite. Bauxite is a mixture of various aluminum oxides.

Aluminum is one of the lighter metals. Its density is about one-third that of iron. The metal has a silvery-gray color. When in contact with air, a thin white coating of aluminum oxide forms at the surface.

Although aluminum is very abundant, it was not widely used until the 20th century — because until then no one had found an efficient, practical way to remove aluminum from its ores. The usual method of roasting or reducing an ore could not be used to extract aluminum from its ores.

✻ In 1886, a 22-year-old student named Charles Hall invented a practical way of removing aluminum from its ore. Hall passed an electric current through a mixture of aluminum oxides (refined bauxite) and cryolite (sodium aluminum fluoride). Heat of the electric current melted the cryolite, which dissolved the aluminum oxide. The electric current then caused the aluminum to separate from its oxide. The molten aluminum was then easily collected. The method of separating a metal from its ore by an electric current is called *electrolysis* (i•ˌlek•ˈträl•ə•səs).

✻ Hall's process, commercialized in 1889, lowered the price of aluminum from $2.00 per pound to 20¢ per pound. Thus, Hall's invention caused aluminum to become one of the world's most widely used metals.

Iron is the most widely used metal. Iron is the second most abundant metal in the crust of the earth. But, commercially it is a more important metal than aluminum. That is because there are *several* ores of iron from which iron can be extracted profitably. Aluminum can still only be commercially extracted from *one* of its ores, *bauxite.*

Refined In this sense, free from impurities.

FIGURE 4-1 Molten iron rushes from a blast furnace through canals (clay-lined runners) and drops into a "submarine" below. Most of this iron will be transferred to another kind of oxygen furnace for refinement into steel. Slag is tapped from a blast furnace several times during the process of reducing the ore. The slag then passes in an opposite direction into huge pots. The pots are loaded on railroad cars for delivery to the slag dump. (*Bethlehem Steel Corp.*)

Iron is found combined with many other elements in ores of the earth's crust. There are several important ores from which iron is removed. Two of the high-grade ores are oxides of iron, *hematite* (Fe_2O_3) and *magnetite* (Fe_3O_4). A low-grade ore of iron is *taconite*. Taconite is a rock in which are mixed small amounts of hematite and magnetite.

Iron can be removed from these oxide ores by reduction. The reduction takes place in a blast furnace. The iron ore is placed in the blast furnace along with coke and limestone. The coke serves two purposes in the blast furnace. It is burned to provide the heat needed for the reduction of the ore. The coke also provides the carbon that is the reducing agent.

Many impurities may be mixed in with the iron ore. So, limestone is placed in the blast furnace to remove most of these impurities. The limestone reacts with the impurities to form a *slag*.

The melted iron from the reduced ore settles to the bottom of the blast furnace. The slag floats on top of the melted iron. The slag is then skimmed off the top. The melted iron is then drawn from the bottom of the blast furnace (FIG. 4-1).

Iron from the blast furnace, containing about 6 to 8 percent impurities, is cast into blocks. Since these blocks are called pigs, this iron is often called *pig iron* (FIG. 4-2).

Some pig iron is remelted and cast into useful objects. It is then called cast iron. Cast iron is not a high quality iron. It tends to crack when put under sudden stresses. This iron would not be a good material for making such things as nails, hammer heads, or automobile bumpers.

Much of the pig iron from the blast furnace is used to produce steel. Pig iron is placed in a second type of furnace. Scrap steel is also placed in this furnace. The pig iron and scrap steel are melted together.

Slag The wastes from the melting of metals or reduction of ores.

In this second furnace, more impurities are removed from the iron. An almost pure iron remains. Small, but carefully measured amounts of carbon are mixed with the iron. Other metals, such as tungsten, chromium and manganese may also be added. The alloy produced from this mixture is called *steel.*

The amount of carbon or other metal mixed with the iron determines the properties the steel will have — that is, how hard, malleable and ductile the steel will be. The mixture is made according to the properties desired, which are determined by what the steel will be used for.

Calcium is the third most abundant metal in the earth's crust. It has a silver-white color. Calcium is softer than aluminum. **It is a very active metal.** It is said to be active because it combines easily with other elements. One very common compound of calcium is calcium carbonate ($CaCO_3$). Calcium carbonate is the main substance found in marble and limestone.

Another important compound of calcium is *calcium chloride* ($CaCl_2$). Calcium chloride is a type of salt. It is often used as a drying agent to absorb moisture from the air. Bags or small packages of calcium chloride are placed in areas and containers where moist air is not desired.

Some copper is found in the earth's crust as a native metal. Prehistoric people used this pure copper to form tools and weapons. It was probably one of the first metals to be used by prehistoric people. **Today copper is a widely used metal.**

Most copper is found combined with other elements in ores. Copper is usually separated from its ores by *electrolysis.*

Copper is relatively soft, and is very

FIGURE 4-2 This man is dipping a sample of molten pig iron from the stream rushing from the blast furnace. Samples are taken three different times during each blast furnace operation for spectrographic and chemical analysis. Since the temperature of the molten iron is about 1,450°C, the man wears protective clothing. *(Bethlehem Steel Corp.)*

ductile and malleable. It is a yellowish-red metal. Artisans (ˈärt·ə·zənz) find copper easy to work with. As a result, copper is used for many decorative purposes, including the manufacture of jewelry.

Much copper goes into the building of our homes. The electrical wiring in our homes is often made of copper. Copper

Artisan A trained or skilled workman.
Prehistoric Belonging to the period before history was written down.

tubing is also used in the plumbing of our homes. It can also be used as a roofing material.

Copper is often alloyed with other metals. The purpose of alloying it is usually to increase its hardness. Alloys of copper include brass, bronze, pewter and Monel.

Sodium is the sixth most abundant element in the earth's crust. It is a silvery-gray metal that is even softer than calcium. Sodium is *so active* that it is always found combined with other elements in nature.

Since it is so active, pure sodium must be stored in a special way. It cannot be stored in the open air. It would react quickly with the oxygen in the air. Sodium also reacts very quickly when it comes in contact with water. Sodium is often stored in containers of kerosene to prevent these reactions.

Sodium is used in producing certain additives for gasoline. These additives are used to prevent automobile engines from "knocking." Another use for sodium is in atomic power plants. Liquid sodium is used to carry heat energy from the nuclear reactor.

The most useful compound of sodium is *sodium chloride* (NaCl). Sodium chloride is common table salt. It is the salt used in many water-softening devices. Sodium chloride is also used to melt ice on streets and sidewalks.

Metals are essential to our life style. But, as you have noted, metals occur most frequently as compounds. In these compounds, you see that metals frequently are combined with nonmetals. In the nonliving world of matter there is interrelationship between elements that is somewhat similar to the interrelationship that exists in the living world. **Metals interact to build new substances as cells interact to form new organisms.** Cells are the building blocks of living matter. **Elements are the building blocks of nonliving matter.** Could either building block — cell or element — exist without the other?

Of course, an element — metallic or nonmetallic — can be important or valuable even when it is uncombined. Pure gold is valuable. Could you live without oxygen? Without hydrogen?

Silicon is an important nonmetal. Nonmetals normally have more electrons in the outer shells of their atoms than do metals. Nonmetallic atoms tend to gain electrons when they combine with other atoms. That is why they combine readily with metals, which tend to lose electrons (Fig. 4-3). Silicon is the *second most abundant element* in the earth's crust (oxygen is the most abundant). It is found in many different compounds in the crust. It is an important part of the compounds that form many common rocks, clays and sand.

Most sand is made of silicon dioxide (SiO_2). One use of silicon dioxide is in the making of *glass*. Common glass is a mixture of calcium silicate and sodium silicate. These silicates are produced by heating sand with limestone and soda ash.

Other substances can be added to the glass to give it special properties. Boron may be added when a heat-resistant glass is made. Lead oxide may be added to produce the high quality glass used in lenses.

Sulfur is a yellow-colored nonmetal. Deposits of pure sulfur can sometimes be

FIGURE 4-3 An atom is said to be stable when it has 8 electrons in its outer orbit. (Exception: If only one orbit, an atom is stable with 2 electrons in its orbit.) (a) If these metals lose their outer electrons when they combine with nonmetals, will the part that remains (an ion) be stable? (b) How many electrons must each of these nonmetals gain to become stable?

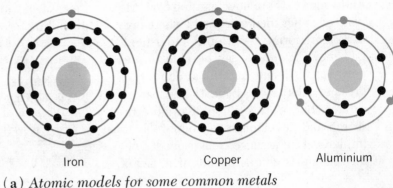

Iron   Copper   Aluminium

(**a**) *Atomic models for some common metals*

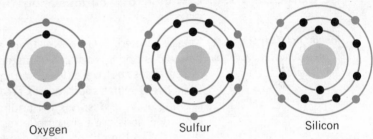

Oxygen   Sulfur   Silicon

(**b**) *Atomic models for some common nonmetals*

found in volcanic areas. Much pure sulfur can also be found underground. In the United States much sulfur is found in Louisiana and Texas.

Sulfur combines readily with many other elements. For this reason, many different compounds of sulfur are also found in nature. Sulfur is often found in compounds with a metal. Many ores of metals contain sulfur. These ores include zinc sulfide (ZnS), copper iron sulfide (CuFeS$_2$), mercury sulfide (HgS), and lead sulfide (PbS).

Sulfur is an ingredient of various medicinal drugs. Sulfur is also used in manufacturing soaps, paints, fertilizers, paper, rubber, gunpowder and matches. Sulfur is used in manufacturing many chemicals. And one product, *sulfuric acid* (H$_2$SO$_4$), is the acid most-widely used in the manu-

facture of other products of the chemical industry.

Scientists disagree as to how long some of the resources of the earth's crust will last. Many scientists say that such metals as copper, iron and aluminum will soon be in short supply. Some say that these metals must be rationed before the end of this century.

Other scientists say that there is an almost unlimited supply of the resources we obtain from the earth's crust. Some say that many materials, such as copper, will soon be obtained from sea water.

Ingredient Something that is a part of any combination such as a mixture or compound.

Earth scientists are now learning from the *plate tectonics* theory how and where ores and oil are formed. With this understanding, they may some day be able to find new ways to locate additional deposits of ore and oil that lie beneath the earth's surface. But, at this time much more research and development of new tools to implement these suggestions is needed before scientists can determine if either of these suggestions can solve the problem of dwindling resources in the earth's crust.

> *Plate tectonics* The theory that states the earth's surface is divided into large, moving sections called plates. Interaction of plates cause geological activity and crustal movement.

BE CURIOUS 4-1: Continue to be curious about the resources of the earth's crust.

Media: newspapers, magazines, newscasts, television

Keep informed. Read newspapers and magazines, listen to newscasts and special programs on TV to find out about the status of resources that were discussed in this section and in the section *Objective 3*. What shortages exist? What shortage will exist? What is being done to assure a continuing supply of these resources? What decisions are being made regarding conservation of resources of the earth's crust? How do you evaluate these decisions? How will they affect your life? life 100 years from now?

OBJECTIVE 4 ACCOMPLISHED? FIND OUT.

1. In what state — combined or uncombined — are most metals found in the earth's crust?
2. What is an ore? ✻ How does a low-grade ore differ from a high-grade ore?
3. What is the science that studies ways to remove a metal from its ore?
4. Name the processes by which a metal can be removed (a) from oxide ores; and (b) from sulfide ores.
✻ 5. Describe and show the chemical reaction for the separation of (a) copper from a copper oxide ore; and (b) of copper from a copper sulfide ore.
6. Which metal is the most abundant in the earth's crust?
7. Why was aluminum not widely used before 1866?
✻ 8. Describe the process invented by Charles Hall.
9. Give one reason why iron is more widely used than aluminum.
10. Why are coke and limestone placed in a blast furnace?
✻ 11. What is steel?

12. Name and compare two very active metals found in the earth's crust.

13. What is the second most abundant element in the earth's crust?

14. State whether each of the following elements is a metal or nonmetal; iron, aluminum, copper, calcium, sodium, silicon and sulfur.

✷ 15. Name the elements found in each of the following substances: calcium carbonate, table salt and common glass.

5 | THE GASEOUS ELEMENTS

YOUR OBJECTIVE: To investigate the properties of some common gases and of some common compounds of these gases, and also the properties of neon and helium.

Nine different elements are found in unpolluted dry air. The elements are found in the gaseous state as elements or compounds. The gases nitrogen and oxygen are very abundant in air. The others — carbon, hydrogen, argon, neon, helium, krypton, and xenon — are very scarce.

Nitrogen is the most abundant gas found in the air. It makes up nearly four-fifths of the air. The nitrogen in the air is found as a diatomic molecule: a molecule made up of two identical atoms (N_2). The diatomic molecule of nitrogen does not react readily with other elements.

At room temperature, nitrogen is a colorless and odorless gas. Its temperature must be lowered to about $-196°$ Celsius before it will turn into a liquid. Nitrogen will turn into a solid at about $-210°$ Celsius.

Many nitrogen compounds have important uses. *Sodium nitrate* ($NaNO_3$) is

one such compound. Another name for sodium nitrate is Chile saltpeter. Sodium nitrate is an ingredient of many fertilizers. *Potassium nitrate* (KNO_3), often called saltpeter, is used in manufacturing explosives. It is also used as a food preservative. *Nitric acid* (HNO_3) is one of the most important acids used in the laboratory and in manufacturing processes. Nitric acid is used in making dynamite, nitroglycerin, and T.N.T. It is also used in manufacturing dyes, plastics and fertilizers. The compound *ammonia* (NH_3) is a colorless gas at room temperature. The gas has a very strong odor. Ammonia dissolved in water is used for cleaning purposes.

Oxygen is a colorless and odorless gas at room temperature. **Pure oxygen makes up about one-fifth of air.** Land plants and animals need this oxygen in order to sur-

Nitroglycerin A heavy oily, explosive liquid used chiefly in making dynamite and in medicine to dilate (widen) the cavities of blood vessels.

FIGURE 5-1 Dark glasses protect the eyes of this welder from the brightness of the very hot, oxygen-fed flame. *(Andrew Sachs for The New York Times)*

vive. Pure oxygen is also found dissolved in water. Water plants and animals need this oxygen.

Most of the oxygen in air appears as a *diatomic gas* (O_2). Large amounts of energy can change O_2 to O_3. *Ozone* is the name given to O_3. Diatomic oxygen changes to ozone when electrical energy is added to the atmosphere during lightning storms.

Oxygen combines very easily with most other elements. In air, it is found combined with carbon as carbon dioxide and carbon monoxide, and with hydrogen as water vapor.

Pure oxygen can be obtained by passing an electric current through water. The electrical energy separates the oxy-

gen and hydrogen atoms in the water molecules.

Another method of obtaining pure oxygen can also be used. This method involves the cooling of air until it becomes a liquid. Liquid air is obtained by cooling the air to about $-200°$ Celsius. The liquid air is then allowed to warm up a few degrees. When this happens, the nitrogen in the liquid boils off. Almost pure oxygen is left behind as a liquid.

Pure oxygen is usually stored in metal tanks. Sometimes the oxygen is mixed with other gases to obtain certain desired properties. The oxygen from these tanks may be used to sustain life.

Oxygen tanks are carried along by skin divers and astronauts. Tanks of oxygen supply the oxygen to the crew and passengers in high flying aircraft. Tanks are also used to supply oxygen to hospital patients.

Industry also has many uses for oxygen. Oxygen tanks are used when welding with a "gas torch" (FIG. 5-1). Pure oxygen also provides the high temperatures for the fires used in the metallurgy of some metals.

The fuel in space rockets must burn in outer space where there is no oxygen. Oxygen is needed for the fuel to burn. For this purpose, oxygen is carried in the liquid state by rockets.

About one-half of one percent of "normal" air is made up of carbon dioxide. Carbon dioxide is not an element, but is a compound of carbon and oxygen (CO_2). It is a colorless and odorless gas at room temperature.

Green plants use carbon dioxide from air to carry out life activities. Carbon dioxide is put into the air by the breathing process of animals. It is also given off as dead plants and animals decay. Large amounts of carbon dioxide are put into the air by the burning of such fuels as coal, fuel oil, gasoline and wood.

The "fizz" in some soft drinks is made by dissolving carbon dioxide in the drink. "Dry ice" is made by cooling and compressing carbon dioxide. "Dry ice" is often used to keep medicines and foods cold during shipment.

Hydrogen is usually found on the earth or in the earth's crust in compounds. Almost no free hydrogen appears in the earth's atmosphere. Most hydrogen found in the atmosphere is combined with oxygen to form water vapor. Hydrogen combines with oxygen to form all of the waters of the earth. Most common fuels are compounds containing hydrogen. Hy-

drogen is also found in the compounds that make up all living things.

Hydrogen is the element found in largest quantity in the sun and other stars. **Scientists estimate that hydrogen makes up about three-fourths of the total mass in the whole universe.**

Hydrogen is the lightest element. It is a colorless, odorless gas at room temperature. It burns with nearly a colorless flame. When it burns, it combines with oxygen to form water.

Because of its lightness, hydrogen gas is often used to fill balloons. Liquid hydrogen is used as a fuel in rockets. When using hydrogen, it must be handled with care. Hydrogen is very explosive.

Like oxygen, hydrogen can be removed from water by passing an electric current through it. Hydrogen can also be obtained by the reaction of an acid with a

BE CURIOUS 5-1:   Find out what percent of "normal" air is oxygen.

Steel wool, water
Test tube, beaker
Grease crayon or
   adhesive tape
Centimeter ruler

Moisten some steel wool with water. Place the moistened steel wool in the bottom of a test tube. Place the test tube upside down in a beaker of water (see the figure at the left). Place a mark, or tiny strip of adhesive tape, on the test tube even with the water level inside the test tube. Allow the setup to remain undisturbed for several days.

Observe the new water level inside the test tube once each day. Place a mark at the water level each day. Continue your observations until the water level no longer rises in the test tube.

Recall your investigation of rusting. Why did the water level rise in the test tube? Why did it stop rising? Measure and compare the size of the air spaces of the first and last observations. What percent of the air space in the test tube was replaced by water? What happened to the space between the marks as the days went by? Explain. How does the percent of air space replaced by water in the test tube compare with the percent of oxygen found in air?

Scientists build upon the discoveries of others. A French scientist, Antoine Lavoisier, built upon the discovery of an English scientist, Joseph Priestley, to gain a better understanding of oxygen and air.

In 1774, Joseph Priestley heated an ore of mercury. The heating caused silver drops of mercury to be released from the red ore. Priestley found that a wooden stick burned brightly when brought near the heated ore. He suspected that a gas released by the ore caused this to happen. He collected this gas in a bottle and did further experiments with it.

Priestly burned a candle in the gas he collected. He found that the candle burned faster and brighter in this gas. He placed a mouse in a bottle of this gas. The gas caused the mouse to become very active. Priestly called this gas "good air." He found also that when the mouse used up the "good air" he died, but that green plants could restore "good air."

The French scientist, Antoine Lavoisier, heard about Priestley's discovery. Lavoisier decided to do more experiments with the "good air" discovered by Priestley. Lavoisier's experiments showed that the "good air" was one of the gases found in the earth's atmosphere. He was also able to explain combustion by analyzing what occurred when "good air" combined with a fuel. Lavoisier gave the name oxygen to "good air," and determined the part oxygen played in the respiration of plants and animals. Respiration is the process by which living things exchange and make use of carbon dioxide and oxygen.

---

metal. This reaction causes hydrogen bubbles to be released through the acid. The following equations show such a reaction.

Zinc + Sulfuric acid ⟶ Zinc sulfate
+ Hydrogen gas

$$Zn + H_2SO_4 \longrightarrow ZnSO_4 + H_2\uparrow$$

Argon makes up a little less than one percent of the atmosphere. It is called an *inert* or "lazy" gas because it does not combine easily with other elements. Argon gas is used inside the glass enclosure of light bulbs. The gas will not react with the hot filament in the bulb. It prevents the filament from burning up when it gets hot.

Other inert gases include neon and helium. Neon glows red when an electrical discharge passes through it. The red glow of neon is often seen in advertising signs.

Helium is the second lightest element. Like hydrogen, it is often used in filling balloons. Unlike hydrogen, it is not explosive. Large amounts of helium are not found on the earth. It is much more plentiful on the sun. In fact, the element was discovered as scientists studied the sun's atmosphere.

BE CURIOUS 5-2: **Investigate some properties of carbon dioxide.**

Part A

Baking powder
2 beakers
Teaspoon
Vinegar
Splint
Match

Place about 2 teaspoons of baking powder in the bottom of a small beaker. Pour about 2 teaspoons of vinegar on the baking soda. This reaction causes carbon dioxide gas to be released. How can you tell that a gas is formed? Hold a lighted match or splint in the beaker (see figure (a) at the right). What is the result? What are your conclusions?

(a)

Part B

Carbon dioxide is heavier than air. Therefore, it can be poured like water. Carefully pour the carbon dioxide that you produced into a second beaker as shown in figure (b) at the right. Hold a lighted match or splint in the second beaker. Were you able to pour the carbon dioxide? Explain.

(b)

✳ **Part C**

On the basis of what you have observed, suggest one use for carbon dioxide. Explain.

OBJECTIVE 5
ACCOMPLISHED?
FIND OUT.

1. What are the two most abundant gases found in the air? In what percents are they found?
2. What is a diatomic molecule?
✳ 3. What is "good air"? Describe its discovery.
4. Name the elements found in each of the following substances: ammonia, water, carbon dioxide, saltpeter and nitric acid.
5. What is the most abundant element found in the universe?
6. What is an inert gas? Name three inert gases. ✳ Describe the properties of each.
7. Why is helium more desirable to use in filling balloons than hydrogen?
✳ 8. Explain why it would not be desirable to fill a light bulb with oxygen or hydrogen.

YOUR OBJECTIVE: To identify pollutants commonly found in air and to find out how each forms and what can be done to reduce or eliminate the pollutant.

Maintaining a supply of clean, unpolluted air is necessary to life on earth.

Plants and animals need supplies of certain pure gases to maintain their life activities. Plants and animals must also be protected from harmful gases and other pollutants in order to maintain their normal life activities.

Natural events can cause the air to become polluted. The eruption of a volcano

FIGURE 6-1   Smoke from this forest fire in Ochoco National Forest, Oregon blackens the sky. About 100 acres of timber burned before the fire was brought under control. Lightning often causes forest fires. How might "people" activities cause a forest fire? *(USDA from Monkmeyer)*

adds sulfur to the atmosphere. Forest fires put smoke particles and carbon monoxide (CO) into the atmosphere (FIG. 6-1). Decaying vegetation adds methane (CH_4) to the air. Nitrogen oxides are formed when lightning passes through the air.

Natural events also tend to clean the air of impurities. Rain and snow wash many pollutants from the air. Gravity causes smoke particles to settle to the ground after a period of time. Chemical reactions may occur in air that slowly change some pollutants to harmless substances.

People's activities can also cause the air to become polluted. The pollutants produced by such activities are often concentrated in a small area, such as in a city. As a result, the pollutants may build up to a dangerous level.

Various types of dust, particulate matter, can be found in the air. Some of the particulate matter — wind-blown soil, volcanic dust, and salt particles — comes from natural sources (FIG. 6-2). Other particulate matter in air is the result of "people" activities. Ash and soot from chimneys, or fine dust from mining and industrial operations are examples of particulate pollution caused by "people" activity.

The pollution of air cannot always be seen. Poisonous, but invisible gases may be escaping from the smokestack of a factory. Invisible poisonous gases escape from automobile exhausts. On the other hand, what appears to be pollution may not be. For example, the thick white "smoke" escaping from some factory smokestack may only be harmless water vapor.

Some pollutants in the air are gases.

FIGURE 6-2 Trash is burned at a dump in Washington, D.C. This happens in dumps from coast–to–coast in our country. Try to estimate how many dump fires may be burning in the United States in one day. *(Paul Conklin from Monkmeyer)*

The gases that present the greatest pollution problems are carbon monoxide, sulfur oxides, nitrogen oxides and hydrocarbons.

Carbon monoxide (CO) is a colorless, odorless gas. When inhaled, it reduces the amount of oxygen in the bloodstream. Shortage of oxygen in the bloodstream can cause brain damage and death. Carbon monoxide, like many other pollutants, can be especially harmful to older people and persons in poor health.

Find out what is done by your state to eliminate undesirable car-bus-truck gas emission.

Community resources Use community resources, such as the police department or motor vehicle inspection station, to find out (a) what standards have been established; (b) what tests are carried out on motor vehicles to find out whether each vehicle meets state emission standards; and (c) what is the penalty for driving a vehicle that fails inspection.

Carbon monoxide enters the air in large amounts due to the incomplete combustion of fuels. Much carbon monoxide pollution comes from the incomplete combustion of gasoline in the automobile engine. This occurs because of a shortage of oxygen in the gas-air mixture in the engine.

The amount of carbon monoxide produced by an engine can be reduced. Increasing the amount of air in the gas-air mixture will reduce the amount of carbon monoxide produced. Burning the fuel at a higher temperature also results in more complete combustion and reduces the amount of carbon monoxide produced. But, these solutions are not as simple as they seem. They reduce the amount of carbon monoxide, but they can make the engine run rougher and increase the amount of other pollutants released.

Recent improvements in the automobile engine are designed to make combustion more complete. Pollution control devices attached to engines also remove much of the carbon monoxide that is produced.

Sulfur dioxide (SO_2) and sulfur trioxide (SO_3) are two dangerous air pollutants. Sulfur dioxide is a poisonous gas. Much of it gets into the air when coal and oil with a high sulfur content is burned. Smaller amounts enter the air from the processing of ores containing sulfur.

In sunlight, sulfur dioxide reacts with oxygen in air. A reaction such as this, which takes place due to the energy of sunlight is called a *photochemical reaction*. This photochemical reaction produces sulfur trioxide. Sulfur trioxide, in turn can react with water vapor in the air. This reaction forms sulfuric acid (H_2SO_4). The chemical equations for the two reactions can be written as follows:

In sunlight

$$SO_2 + O_2 \longrightarrow SO_3$$

$$SO_3 + H_2O \longrightarrow H_2SO_4$$

Sulfuric acid is a very strong acid. Besides being harmful to the human body, the acid in the air harms many nonliving things. The acid reacts with many paints, fabrics, metals and certain types of building stones. Sulfuric acid causes these materials to decay faster.

The amount of sulfur oxides in air can be reduced by burning fuels that are low in sulfur content. Most low-grade coals are high in sulfur content. But, the demand for energy and the shortage of fuels at the present time require that some fuels with high sulfur content be used.

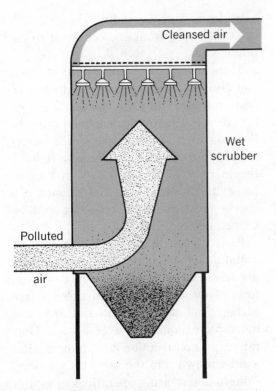

FIGURE 6-3 Diagram of a wet scrubber. The scrubber duplicates the natural process of rainfall and washes particles out of smoke or gas. *(Manufacturing Chemists Association)*

FIGURE 6-4 Emission of unburned or partly burned fuels contributed to the formation of the smog that hangs over Chicago. *(Mimi Forsythe from Monkmeyer)*

When high-sulfur fuels are burned, various methods can be used to remove the sulfur oxides from the exhaust. One method involves passing the exhaust through a wet *scrubber* (FIG. 6-3). Since both oxides of sulfur dissolve in water, the shower removes much of these harmful gases from the exhaust.

Under ordinary conditions, nitrogen and oxygen in air do not combine with each other. However, they do combine at very high temperatures. These temperatures occur in high temperature furnaces used in power plants and some manufacturing operations. Very high tempera-

tures are also reached during combustion within the automobile engine.

Under very high temperatures nitrogen and oxygen combine to form nitrogen oxide (NO). When the nitrogen oxide is released into the air, it can form nitrogen dioxide (NO_2). Nitrogen dioxide forms during a photochemical reaction between the nitrogen oxide and other substances in air.

Nitrogen dioxide is a brownish gas. It can be washed away by rainfall or scattered by the wind. If this does not occur, the nitrogen dioxide helps to form a *smog* (FIG. 6-4). Smog can weaken rubber and fabrics. It can kill plants. Smog irritates the eyes and lungs of people, and can cause death to persons with respiratory, cardiac, and cardiovascular diseases.

The emission of nitrogen oxides from auto engines and other sources can be reduced. Combustion at a lower temperature reduces the amount of nitrogen oxides produced. However, lower temperatures can cause an increase in carbon monoxide. Exhausts containing nitrogen oxides can be passed through a catalytic converter. The catalytic converter changes the nitrogen oxides back into the elements oxygen and nitrogen.

Hydrocarbons are compounds that contain only hydrogen and carbon. There are many hydrocarbons. Hydrocarbons of such substances as natural gas, gasoline and kerosene can pollute air. These hydrocarbons can get into air when liquid hydrocarbons evaporate. They can also get into the air when fuels are only partly burned.

Many chemical reactions can occur when hydrocarbons are in the air. Some form *particulates*. Some hydrocarbons form *smog*. **Hydrocarbons and nitrogen oxides are the main cause of smog.**

One way to reduce the amount of hydrocarbons that enter the air is to prevent their evaporation from liquid hydrocarbons. Storage tanks containing such liquids should be covered. Vapors should be kept from escaping while these liquids are transferred from one container to another.

More complete combustion of fuels can also reduce the amount of hydrocarbons released into air. Better designed furnaces and engines can accomplish this. Burning fuels in an engine for a longer time before exhausting them also reduces the amount of hydrocarbons in the exhaust. Hydrocarbons in the exhaust can also be recycled through the engine to be burned again.

Particulates are the most easily seen air pollutants. **Particulates are very tiny solid particles carried in air.** Particulates come from many sources. Tiny particles of free carbon can be released into the air by the incomplete combustion of a fuel. Dust raised at construction and mining sites can be blown into the air. Fine particles from manufacturing operations can enter the air through chimneys.

Volcanic eruptions send huge clouds of dust into the air. In October of 1974 a volcanic eruption in Guatemala threw a vast dust cloud into the stratosphere. As the cloud moved northward the dust particles caused spectacular sunsets. By early 1975 the sunsets had been seen in widely separated areas: Hawaii, Wyoming, New York, England and France.

Some particulates are very small particles. These small particles can be carried great distances by the wind. Other particulates are much larger. The larger particles may settle to the ground quickly and are often called *dustfall*.

The total effect of particulates in the air it not known. Some may have a harmful effect on the respiratory systems of living things. Large amounts of particu-

Cardiac Involving the heart.
Cardiovascular Involving the heart and blood vessels.
Evaporation The process by which a liquid changes into a gas.

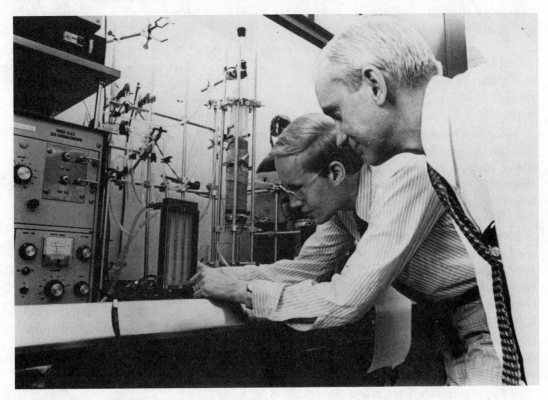

Rudie J. H. Voorhoeve and Joseph P. Remeika of Bell Laboratories test some new catalytic materials.

Scientists are experimenting with many different ways to reduce air pollution. Many are working on ways to reduce the pollution caused by automobile engines. Some scientists at the Bell Laboratories at Murray Hill, N.J. are working on such a project. You may wonder why a project such as reducing auto engine pollution is being carried out in a research laboratory of a telephone company. It is because the Bell system has one of the largest motor vehicle fleets in the nation and is concerned with the fleet's role in improving air quality. Also, they are concerned with the protection of sensitive telephone equipment from the effects of air pollution, some of which comes from autos.

These scientists are experimenting with some low-cost materials that can be used in catalytic converters. The converters are containers attached to the automobile's exhaust system. The converters can cause pollutants to be changed into harmless gases.

The scientists are trying out different catalysts in the converters. Catalysts are materials that cause desired chemical reactions to take place in the converters: in this instance reactions that change the pollutants to harmless gases. The scientists are trying to find long-lived, inexpensive catalysts that will do an efficient job in eliminating the pollutants.

late matter in the air may change the amount of rainfall in an area.

It is also thought that large amounts of particulate matter in the air could lower the temperature of the earth. The particulates could prevent some of the sun's energy from reaching and heating the earth.

The amount of particulates in the air can be reduced in many different ways. **Air containing particulates can be passed through a filter. Particulates too large to pass through the filter are removed from the air.**

Electrostatic precipitators can also be used to remove particulates. First, the particulate matter must be given an electric charge. The air containing the charged particulates flows past electrically charged plates (Fig. 6-5). The negatively charged particles are attracted to the positive plates. The positive particles are attracted to the negative plates. Electrostatic precipitators are often placed in chimneys to remove particulate matter from the escaping gas.

Wet scrubbers are also used to remove particulate matter. The air is passed through a series of showers. The liquid in the shower washes the particulates to the

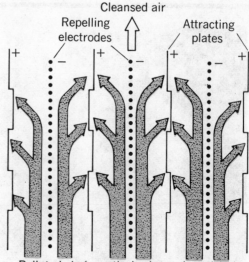

FIGURE 6-5 Diagram of an electrostatic precipitator. A strong electrical charge is placed upon the particles to be removed. The charged particles are attracted to the oppositely charged plates in the chamber. Generally, the plates are cleaned with a *rapping* device that causes the particles to fall off for collection. How does the principle of static electricity explain this operation? (*After a diagram from the Manufacturing Chemists Association*)

bottom of the scrubber. The wet scrubber can also be used to remove some soluble gaseous pollutants from the air.

OBJECTIVE 6
ACCOMPLISHED?
FIND OUT.

1. What is a cause of much carbon monoxide being released into the air?
2. What is a major cause of sulfur oxides in the air?
✻ 3. How is sulfuric acid formed by the sulfur oxides in air?
4. What is a photochemical reaction?
5. Under what conditions will nitrogen and oxygen combine to form nitrogen oxide? Where does this condition exist?
6. What gases are the main cause of smog?

7. What is a particulate air pollutant?
8. Describe one method used to remove particulate pollutants from air.

7 | THE CHEMISTRY OF WATER POLLUTION

YOUR OBJECTIVE: To identify pollutants commonly found in water; to investigate the source of each and what can be done to reduce or eliminate the pollutant.

To stay alive you need clean water. The cells of your body must have water in order to function properly.

About three-quarters of the earth's surface is covered by streams, lakes, seas, oceans. Water is stored in the earth's crust. In most places in the world there is not a severe shortage of water. However, one-third of the earth's surface is desert (Fig. 7-1). And, in many other places there is a severe shortage of fresh water *suitable for drinking* (Fig. 7-1).

What kind of water is suitable for drinking? To be suitable for drinking water must be free of any organisms or other substances that can cause illness or death. The water should also be free of any bad taste or odor. It is also desirable that the water be clear and cool.

Certain amounts and kinds of water pollution are controlled naturally. Suppose that certain materials enter a stream at point A and pollute the stream. Unless water in the stream is moving very rapidly, *solid pollutants* soon *settle* to the bottom. *Bacteria* in the water cause some

of polluting substances to be destroyed. *Sunlight* and *oxygen* in the water destroy some other pollutants. At point B downstream, the water may once again be pure.

FIGURE 7-1 This region of Arizona would still be desert without the water provided by the Salt River Project. (*Monkmeyer*)

On the other hand, water can be so badly polluted that natural events cannot remove the impurities fast enough. Suppose that untreated sewage from a fair-sized community was discharged into a stream. Natural events probably could not purify that water before it reached the next community a few miles downstream. If the next community depends on the stream for its drinking water, steps must be taken to purify the polluted water. **Water pollutants include particulate matter, microorganisms, dissolved gases and other chemical substances.**

Particulate pollutants in water are called suspended solids. These are materials that are not dissolved by the water. They are carried in the water by its motion. When water motion is slight, or water becomes still, suspended solids settle to the bottom. Generally, the heavier the particles the more quickly they tend to settle.

Suspended solids may often consist of *soil* that has been washed into the water. They may also come from *untreated sewage.* Sometimes they come from the *tailings* from mine operations.

Suspended solids can be seen in water. **The particles block out sunlight and prevent it from reaching underwater plants.** Without sunlight, the plants are not able to release gaseous oxygen into the water. Without this oxygen, plant and animal life cannot live underwater.

Suspended solids can be removed from water by *filtering* it. They can also be removed by allowing them to *settle* to the bottom in still water. Sometimes it takes a long time for them to settle.

Suspended solids settle faster, when a chemical called a coagulant is added to water. The coagulant causes the small suspended solids to cluster together. This large cluster of solids quickly settles to the bottom.

Many polluted waters contain large numbers of harmful microorganisms. These microorganisms, or germs can cause disease or death. Untreated sewage and manufacturing processes add these organisms to the water. In farming areas, runoff may carry livestock wastes to streams and cause this type of pollution.

Chlorine is used to kill germs in city water and sewage systems. Small amounts of chlorine are added to the water to kill the germs in drinking water. Larger amounts are put in the water used in swimming pools.

Many nonliving substances in water can also be very harmful. Substances such as arsenic, mercury and cyanide can cause death if consumed. They are said to be *toxic.* Some toxic substances kill fish and other water life (FIG. 7-2). Some are stored by fish and other water life and passed on to anyone who may eat them.

Microorganism A very small living substance that cannot be seen without magnification.

Tailings The remains of an ore after the useful element has been removed.
Filter A device or porous substance (substance with holes) used as a strainer.

Runoff The part of rainfall that is not absorbed directly by the soil upon which it falls, but drains off in small streams to other areas (frequently larger streams or ponds).

Toxic substances may come from pesticides that have been washed into the water. Some may come from metal processing wastes. Others may come from the wastes from the manufacture of paper and chemicals.

Some waste chemicals put into water cause the dissolved oxygen in the water to be used up as these waste chemicals undergo chemical reactions in the water. The amount of dissolved oxygen needed for this breaking-down process is called the *biochemical oxygen demand* (BOD).

Bacteria in water aid in breaking down certain organic (carbon containing) waste material. Materials that can be broken down this way by chemical reactions are said to be *biodegradable*. Oxygen is *used up* as these organic materials are broken down. Organic wastes may be put into the water by home sewage. Some industries, such as food processing plants add to this type of waste.

Certain inorganic chemicals such as sulfides, iron, and ammonia also use up oxygen in water as they react with chemicals in the water. These inorganic chemicals may be put into the water by some manufacturing processes. Ammonia is also an ingredient of many household cleansers and can enter water supplies with home sewage.

The amount of dissolved oxygen can also be reduced when too many nutrients are added to the water. Some compounds

FIGURE 7-2 Waves wash ashore fish killed by toxic substances in Lake Michigan. *(Hays from Monkmeyer)*

of nitrogen and phosphorous are nutrients commonly found in water. Nutrients cause water to be over-fertile. As a result, undesirable weeds and algae in the water grow very rapidly. This can cause a lake or river to become choked with growth (FIG. 7-3). When the plants and algae die,

Algae Simple plants, some being made up of only one cell.

BE CURIOUS 7-1: **Investigate the water supply system and sewage treatment system of your community.**

Community resources Use any available sources to find information about water supply and sewage treatment and disposal in your community.

large amounts of oxygen from the water are used up in the chemical reactions of the decay process.

Nutrients that make the water over-fertile come from many sources. The sources include home sewage and industrial waste. Fertilizers washed into streams from farmlands also add nutrients to streams, ponds and lakes.

The oxygen found in water comes from two sources. *Water plants* add oxygen to water as they carry on photosynthesis. Oxygen enters water *from air.* Gaseous oxygen from air enters bodies of water at the surface of the water by dissolving in the water. Whenever the surface of the water is disturbed, the surface area exposed to air becomes greater. A greater surface area allows more oxygen to enter the water. For instance, water running over waterfalls or rapids picks up more oxygen than does water of the stream that leads to the falls or rapids. In water treatment plants oxygen is added to water by spraying the water into the air. This process is called *aeration*.

Warm water can hold less dissolved oxygen than can cold water. Therefore, a change in water temperature can affect the oxygen content of water. A change in temperature and in oxygen content can have serious effects on plants and animals living in water.

The undesirable changing of temperature of a body of water is called thermal pollution. Thermal pollution occurs when large amounts of warm water are discharged into a lake or stream from industrial processes. Much thermal pollution results from discharges from electrical power plants.

Thermal pollution can be greatly reduced by cooling warm discharge water before allowing it to enter a lake or stream. Various methods are being tried to cool discharge water. All of these methods use air as a coolant. The heat from the warm water is lost to the air.

OBJECTIVE 7
ACCOMPLISHED?
FIND OUT.

1. What kind of water is suitable for drinking?
2. What kinds of pollutants are found in water?
3. What methods might be used to remove suspended solids from water?
4. Why is a coagulant added to polluted water?
5. What method is commonly used to kill germs in city water supplies?
6. What is meant by biodegradable material?
7. Name two common nutrients that can cause water to become over-fertile.
8. What is the relationship between the amount of dissolved oxygen in water and the temperature of water?
9. What is thermal pollution?
✵ 10. Explain why and how the amount of oxygen found in water varies?

A substance can be described by its chemical and physical properties. The properties of a substance determine what it is used for. In a physical change physical properties of a substance can change, but no new substance is formed. The chemical properties of a substance describe how it interacts with other substances. In a chemical change at least one new substance is formed. Chemical changes occur when substances combine or break down to form new substances.

In a physical change and an ordinary chemical change, mass is conserved. Every chemical reaction involves energy. In exothermic reactions energy is released. In endothermic reactions energy must be added or the reaction will not take place. A chemical equation shows how the various atoms are arranged before and after a chemical reaction. A chemical formula shows what a substance is made of.

Elements, compounds, and mixtures found in the earth's crust are necessary resources for an industrialized nation. Most resources of the earth's crust — such as fuels — are nonrenewable. Some — such as metals — are nonrenewable, but can be recycled. A renewable resource is one that can be replaced naturally in a reasonable amount of time. A nonrenewable resource is one that cannot be replaced naturally in a reasonable amount of time. To use resources wisely, much careful planning (conservation) is necessary. In an industrialized nation fuels and metals are the most important and essential resources of the earth's crust.

Fuels are substances that produce large amounts of heat when burned rapidly. Combustion of fuel is an oxidation reaction in which oxygen of the air combines with carbon or hydrogen of the fuel. The temperature at which a fuel begins to burn is the kindling temperature of the fuel. When the oxygen supply is low during burning, incomplete combustion of a fuel occurs releasing pollutants.

Common fuels include coal, oil, gasoline, and natural gas. In this country, there is a fuel-energy crisis. An immediate answer to the crisis seems to be to return to coal and to draw upon the resources of low-grade coal previously unused; and to continue to develop an efficient plant operation to produce methane by coal gassification. The fuel of the future may well be hydrogen.

Metals are useful because as a group they have desirable physical properties. They are hard, malleable, ductile, have high tensile strength and high melting points, and conduct heat and electricity well. Chemically they react readily with each other and other elements to form a variety of substances. Metals combine readily with gases of the air causing corrosion of the metal. This is an undesirable property.

Alloys are combinations of metals and other elements. Most metals found in the earth's crust are not in the pure state. Compounds of metals that occur naturally in the earth's crust are called minerals. Rocks are mixtures of minerals. A deposit is a concentration of one or more minerals found in a relatively small area of the earth's crust. An ore is a deposit from which metal can be removed profitably. The ores of many metals are oxides or sulfides. A metal can be removed from an oxide by the process of reduction; from a sulfide by the process of roasting followed by reduction.

Among other elements of the earth's crust that are particularly useful are aluminum, iron, calcium, copper, and sodium. Silicon and sulfur are important nonmetals of the earth's crust.

The gases hydrogen, oxygen, and carbon dioxide are essential to life on earth. Unlike metals and fuels, these elements are in abundant supply in the air and waters of the earth and are recycled naturally. The problem with regard to these resources is pollution. Our life style pollutes the air with such substances as carbon monoxide, sulfur, nitrogen oxides, hydrocarbons and particulates. Our life style pollutes the water with suspended solids, disease-carrying microorganisms, poisonous substances, and oxygen-demanding wastes and heat.

UNIT OBJECTIVES ACCOMPLISHED? FIND OUT.

PART A   Match the numbered phrases 1–10 with the lettered terms.

1. The total mass remains the same when a physical or ordinary chemical change takes place.
2. A chemical reaction in which energy is given off.
3. The process by which different compounds are separated from petroleum.
4. Small solid particles found in the air.
5. The amount of dissolved oxygen needed to "digest" wastes in water.
6. An element that tends to lose electrons when it combines.
7. The tarnishing of a metal by its reaction with some gas in the air.

a. active
b. endothermic
c. biochemical oxygen demand
d. conservation of mass
e. corrosion
f. distillation
g. exothermic reaction
h. metal
i. nonmetal
j. particulate matter
k. steel

8. A term used to indicate that an element combines easily with other elements.
9. An alloy containing iron and other selected metals.
10. An element that tends to gain electrons when it combines.

Part B   Choose your answer carefully.

1. How a substance reacts with other substances is described by its (a) biochemical oxygen demand   (b) tendency to gain electrons (c) physical properties   (d) chemical properties.
2. Which of the following is not an example of a physical change? (a) melting   (b) breaking   (c) bending   (d) rusting
3. All chemical reactions involve   (a) increases in total mass (b) energy changes   (c) the production of a gas   (d) decreases in total mass.
4. PbS $\longrightarrow$ Pb + S is called   (a) a chemical formula   (b) a chemical equation   (c) a word equation   (d) a coefficient.
5. In the equation $CaCO_3$ + heat $\longrightarrow$ CaO + CO_2, $CaCO_3$ is called   (a) an element   (b) a hydrocarbon   (c) a reactant (d) a product.
6. The burning of coal is not an example of   (a) rapid oxidation (b) an exothermic reaction   (c) combustion   (d) an endothermic reaction.
7. A common product of incomplete combustion, which is not found in complete combustion, is   (a) carbon monoxide   (b) carbon dioxide   (c) oxygen   (d) water.
8. When wood is heated in the absence of oxygen   (a) coke (b) charcoal   (c) slag   (d) steel is produced.
9. A material is said to   (a) be ductile   (b) have a high tensile strength   (c) be malleable   (d) be hard if it can be rolled into a thin sheet without cracking.
10. Corrosion is most damaging to the metal   (a) aluminum (b) silver   (c) copper   (d) iron.
11. A blast furnace removes iron from its ore by the process of (a) roasting   (b) filtering   (c) reduction   (d) electrolysis.
12. The metal   (a) sulfur   (b) iron   (c) gold   (d) aluminum can be found in the earth's crust as a native metal.

13. A   (a) nonrenewable   (b) recycled   (c) reduced   (d) renewable material is one that nature cannot replace in a reasonable amount of time.

14. Sulfur oxides from combustion of high-sulfur fuels can be removed from exhausts by   (a) a catalytic converter   (b) a wet scrubber   (c) filtering   (d) settling.

15. Under ordinary conditions   (a) hydrogen and oxygen   (b) carbon and oxygen   (c) sulfur and oxygen   (d) nitrogen and oxygen do not combine in air.

16. The main causes of smog are   (a) hydrocarbons and nitrogen oxides   (b) carbon monoxide and nitric oxides   (c) particulates and carbon monoxide   (d) nitrogen oxides and particulates.

17. In sunlight sulfur dioxide reacts with oxygen to produce sulfur trioxide. Such a reaction is called   (a) exothermic   (b) photochemical   (c) combustion   (d) electrostatic.

18. Suspended solids in water   (a) can be removed by filtering   (b) are called coagulants   (c) are dissolved in water   (d) are not considered to be pollutants.

19. Chlorine is often added to water to   (a) act as a coagulant   (b) kill germs   (c) cause aeration   (d) remove toxic substances.

20. A direct result of thermal pollution is the   (a) reduction of oxygen in water   (b) over-fertilization of the water   (c) growth of toxic substances   (d) formation of suspended solids.

✷ **Part C**   Think about and discuss these questions.

1. Would you or would you not build a bridge of silver across a large river? Explain.

2. Discuss possible short- and long-term solutions to the problem of dwindling fuel supplies. What are the advantages and disadvantages of each suggestion?

3. Explain why a shortage of metals can be more easily solved than a shortage of fuels.

4. Do you consider air or water pollution to be equally serious or is — in your opinion — one more serious than the other? Explain.

MOTION AND FORCES

This big machine is the "whirlwind ride" located at North Hudson Park in North Bergen, N.J. In the upper view you can see it standing still as the prospective riders board the machine, settle into their seats, and prepare to hang on for dear life.

This machine is driven by a large motor that applies a torque, or spinning force, to the drive mechanism. In the lower view you can see the whirlwind ride whirl — like a whirlwind indeed!

1 | WHAT IS MOTION?

YOUR OBJECTIVE: To define motion within a frame of reference; to understand the meaning of speed, velocity and acceleration and how to solve problems dealing with these quantities.

What is motion? At first glance, the answer to this question may seem very simple. Does a book resting on a table top have motion? What if the book and the table are located on an ocean liner crossing the Atlantic? Remember that the Earth is moving around the sun before considering the next question. Is the book moving if it is resting on a table at home?

Scientists describe the motion of an object by *comparing* it to something else. A book resting on a table top is at rest when compared to the table, or to the floor on which the table stands. It is also at rest compared to the room where the table is located. If the table is on a ship which is crossing the ocean, the book is moving compared to a location on land. The table in your home is moving when it is compared to a location on the sun, and so is the book resting on it.

When the motion of an object is compared to something else, that something else is called a frame of reference. In the case of the book resting on the table top on the ocean liner, the book is at rest if the ship is used as the frame of reference. If the land from which the ship is moving is used as the frame of reference, then the book is moving along at the same rate that the ship is moving. If the sun is used as a frame of reference, then the motion of the book is far more complicated to describe.

When describing the motion of an object it is usually best to choose the simplest frame of reference. Suppose someone tells you that an automobile is moving at 50 miles per hour. You take it for granted that the road on which the car is moving is used as the frame of reference. When two people play catch with a baseball the simplest frame of reference to use for describing the motion of the ball is the ground on which they stand.

One characteristic of motion that you are very familiar with is *speed*. When you see an object in motion in some frame of reference you are aware that it is moving either quickly or slowly.

It is often useful to measure how quickly or slowly the object is moving. How would you go about doing this? In order to measure how slow or how fast an object moves, you need to know the *displacement* of the object, and the *time* it took for this displacement to occur.

Speed is defined as the distance an object travels in a certain amount of time. It is measured by comparing units of distance such as feet, miles, meters, or kilometers to units of time, such as seconds or hours. This comparison may be expressed

> *Displacement* As used, the relation between a moving object at any time and its original position.

Try changing your frame of reference.

In this investigation you will be asked to imagine the motion of various objects compared to some frame of reference. You should then try to describe the motion from a different point of view or frame of reference. Describe the motion in each of the following cases using *words* or *drawings*.

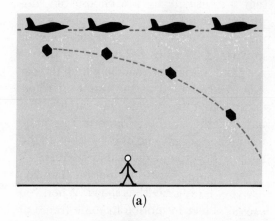

(a)

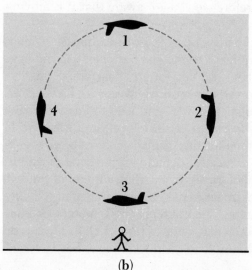

(b)

Part A

Two airplanes are flying next to each other at the same speed. Describe the motion of the one airplane as seen by the pilot of the other plane.

Part B

Figure (a) shows the path of a package dropped from an airplane as seen by someone on the ground. Also describe the path of the falling package as seen by someone in the airplane.

Part C

A person standing on a railroad flat car tosses a rock into the air and catches it on its way down. The flat car is moving along the tracks at a steady speed. How would a person describe the motion of the rock as seen from a point alongside the railroad tracks?

Part D

An ant is standing on a slowly spinning phonograph record. How does the ant see the motion of a lamp bulb on the ceiling above the phonograph?

Part E

A person on the ground sees an airplane flying in a loop as shown in Figure (b). Describe the motion of this person as seen by the pilot.

as *meters per second* or *miles per* hour. If someone tells you that the jet you are flying in is going at 500 miles per hour, it means that if the plane continues to fly at this speed, you will travel a distance of 500 miles in one hour's time. (Unless noted otherwise, speeds are given with the Earth as the frame of reference.)

The speedometer of a car gives a reading of the car's speed for any given moment while it is in motion. On a trip, the speedometer will register many different readings. Cruising along the turnpike it may show readings of 45, 50, or 55 miles per hour. Slowing down for exits from the main highway the readings will more likely be 30, 25, or 15 miles per hour. This sort of speed reading lets you know if you are observing speed limits. **Physicists call this speed instantaneous speed, or speed at any given instant of time.**

Sometimes it is more useful to know what speed you averaged for an entire trip. The speedometer will not give you this information. But it is easy to figure out if you keep a record of the time it takes for the entire trip; and if you keep track of how many miles you travel. This can be read from the mileage gauge on the dashboard. Let us say that in 2 hours you drove 90 miles. **To find the average speed at which you traveled, you divide the distance traveled by the time it took to make the trip.** The average speed of your car during this trip — 90 miles in 2 hours — would be 45 miles per hour.

You can find the average speed easily by using the simple equation

$$average\ speed = \frac{distance}{time}$$

This equation may be abbreviated by using symbols to represent speed, distance and time.

If s_{av} represents average speed, d represents distance and t represents time, then the equation may be written simply as

$$s_{av} = \frac{d}{t},\ d = s_{av}t$$

$$\text{or } t = \frac{d}{s_{av}}$$

In this equation s_{av} is used instead of just s so that the speed at any given moment (speedometer reading) can be distinguished from average speed.

Using what you have just learned, try this problem. An Amtrack train left Chicago for New York at 4:00 pm and arrived in New York City's Penn Station 18 hours later. The distance traveled figured from the mileage gauge in the engine cab was 900 miles. What was the average speed of the train for this trip? If you did your mathematics correctly, the average speed should come out to be 50 miles per hour.

If you are already familiar with the word *velocity*, you may think of it as another word for speed or "how fast." **To a scientist the word velocity means not only how fast something is moving but also in what direction the motion is taking place. Velocity includes both speed and direction.** When you say that a car is going at 50 miles per hour, you are talking about the speed of the car. If you say that a car is going at 50 miles per hour *northward*, you are now talking about the car's velocity. This is an important distinction

FIGURE 1-1 Racing cars accelerating at the start of the 59th Indy 500 race at the Indianapolis Motor Speedway. Gordon Johncock in his Wildcat-Turbo Drake (car 20) leads pole holder A. J. Foyt in his Coyote-Turbo Foyt (car 14). *(Wide World Photos.)*

to make for the following reason. Shortly you will be considering motion where the velocity is *changing*. This change may be in speed, or in direction, or both. A car going at 50 miles per hour along the highway in a direction due north may change this direction as it moves into a curve. Although the car is still moving at a 50 mile-per-hour speed, its velocity changes. You must always keep this in mind when dealing with motion which is not along a straight line.

Another characteristic of motion which you have often experienced while riding in a car is that of *acceleration*. You may refer to this kind of motion as "pick up," meaning the ability of a car to go from a low to a high speed in a short period of time (FIG. 1-1). **The scientist defines acceleration as a change in speed or velocity occurring during a certain amount of time.** This change may be from low to high speed or from high to low speed. In the latter case the word *deceleration* is sometimes used.

✶ Recall for a moment how **speed was defined as a change in distance occurring in a certain amount of time.** Recall also that this relationship between distance and time was expressed in the equation $s = d/t$. Now recall that acceleration was defined as a *change* in speed or velocity occurring during a certain amount of time. The relationship between speed and time in acceleration is similar to the relationship between distance and time in defining speed. So, this relationship can also be simply expressed in an equation. If a is used to represent acceleration and Δv is used to represent a *change* in speed or velocity, and t to represent time, then acceleration can be expressed as

$$a = \frac{\Delta v}{t}$$

✶ You have seen how an object is displaced when it has actually moved. But **sometimes an object appears to be displaced when it is really standing still.**

Δ The Greek letter delta used as a symbol to mean "the change in."

✶ SAMPLE PROBLEM: You are driving an automobile along a straight stretch of highway at 40 miles per hour and you want to pass a truck. As you step harder on the gas pedal, the speed increases from 40 to 60 miles per hour in 5 seconds. What is the acceleration of the car?

Using the equation

$$a = \frac{\Delta v}{t}$$

$$= \frac{(60 - 40) \text{ miles per hour}}{5 \text{ seconds}}$$

$$= \frac{20 \text{ miles per hour per second}}{5}$$

$$= 4 \text{ miles per hour per second}$$

(a)

FIGURE 1-2   This scene of lower Manhattan (New York City) shows the World Trade Center rising above the other buildings. Battery Park is in the foreground. (a) You can see both twin towers of the World Trade Center — a shadow is cast by one tower on the other. (b) You are looking at the same scene from an observation point to the right of your position in (a). Why do you now see only one of the towers? What is this an example of? *(Wide World Photos.)*

This false appearance of motion is called parallax (FIG. 1-2). If you are in a moving vehicle — a car, train, boat — that is passing a group of buildings and telegraph poles, the positions of these objects will appear to have been displaced as you move past them from one point to another point, even though their positions are fixed. This is because you are seeing these things from different points of view.

Astronomers make use of the parallax effect to measure the distance of stars. As the earth moves around the sun there is an apparent change of position between a nearer and a more distant star. Parallax occurs because you observe one star relative to another from two different positions of your observation point. The greater the parallax effect, the greater the distance separating the two stars.

(b)

1. State the simplest frame of reference to use when:
   (a) discussing the motion of a train.
   (b) discussing the motion of the moon around the Earth.
   (c) discussing the motion of the Earth around the sun.
   (d) watching a person dive into a swimming pool aboard a moving ocean liner.

2. What is instantaneous speed? average speed?

3. What is speed? velocity? acceleration?

4. An airplane flew from New York to California in five hours. Find the average speed of the airplane if the distance recorded was 2,700 miles.

✻ 5. What is parallax?

✻ 6. Car A and car B are going along a straight road. Car A is moving at a constant speed of 60 miles per hour. Car B changes its speed from 20 miles per hour to 50 miles per hour in 10 seconds of time. Which car has the greater acceleration? What is the acceleration of that car?

2 | WHAT IS FORCE?

YOUR OBJECTIVE: To learn the scientific definition for force and how force is measured; to recognize friction as a force that can be useful or a nuisance; to understand how forces can result in equilibrium.

As a scientific term, **force is defined as a push or a pull applied to an object.** The action of applying force may also include lifting, twisting, stretching, squeezing and bending.

You are already familiar with the pushing or pulling forces such as pushing a lawn mower or pulling a door open. Other kinds of forces include gravitational force, frictional force, electric and magnetic forces. Force is a very important part of the study of matter, energy and motion. It is important to learn how force is measured and how force affects objects upon which it acts. **Physics is the science that deals with forces and how they affect matter.**

One of the forces you experience most often is *weight*. In the English system of measurement, weight is measured in such units as pounds and ounces. In the English system the units used to measure weight are also used to measure force. However, scientists prefer to use special units for measuring force. One such unit is the *newton* (nt). One newton of force is

> *Weight* The amount of gravity pulling on the mass of an object.

equivalent to about 3 ounces or one-fifth of a pound.

Figure 2-1 shows two stick figures pulling a cart in the *same* direction. Suppose each figure pulls with a force of 20 newtons. The result of these two forces is the same as if the cart were being pulled by only one force of 40 newtons. **When forces act in the same direction, the resulting force is equal to the sum of the individual forces.**

Now suppose two stick figures push a cart in opposite directions (FIG. 2-2). **Because the forces oppose each other, the force pushing on the right cancels a part of the force pushing on the left.** Since the figure on the left is pushing with a force of 30 newtons and the figure on the right is pushing with a force of 10 newtons, but against the figure on the left, you can subtract these forces. Therefore, the result is the same as if the cart were being pushed by a single force of 20 newtons.

Sometimes forces can act in such a way that it appears as if *no* force is acting at all. Two *equal* forces can act in opposite directions (FIG. 2-3). **When two equal forces oppose each other, one force entirely cancels the other force.** In this case the forces are said to be in *equilibrium*. The forces are *balanced* and there is *no* movement of the cart. It appears as if *no* force is acting on the cart. It follows, then, that **an object at rest needs an unbalanced force to set it in motion. If an object is in motion with a constant speed or velocity, it also will need an unbalanced force acting upon it to change its**

FIGURE 2-1 The arrows show the direction of the pulling force on the cart. These two stick figures could be replaced with one that pulls with a force of 40 newtons.

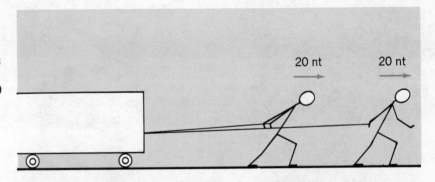

FIGURE 2-2 The arrows show the direction of the push and pull on the cart.

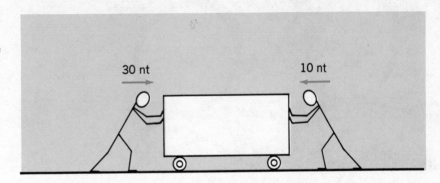

FIGURE 2-3 The arrows show the direction on the pull on cart. Since the two forces are equal and pull against each other, you can subtract them. Therefore 20 nt − 20 nt = 0 nt.

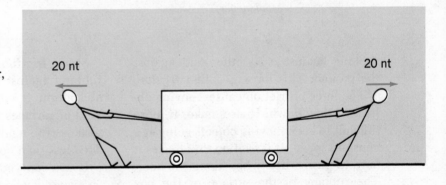

speed or velocity. This is a very important point to understand in studying motion. Remember that **an unbalanced force is needed to change the velocity of a moving object.**

If you stop pedaling a bicycle while riding along level ground, you will notice that the bicycle gradually slows down. Eventually it will come to a stop. There is a force that must be acting to stop the motion of the bicycle (FIG. 2-4). Where does this force come from? You may already know that this force is due to the friction of the moving parts of the bicycle

FIGURE 2-4   What is the force that keeps this bicycle moving forward? *(Hugh Rogers of Monkmeyer)*

rubbing against each other and against the ground. This force is called *friction*.

The force of friction causes moving objects to slow down. It also makes it more difficult to keep moving objects going at a constant speed. It is friction that makes it hard to push a heavy box along the floor. The rubbing of the surface of the box against the surface of the floor results in frictional force which you must overcome in order to move the box.

Constant   The same or un-changing.

Since friction is caused by surfaces rubbing against each other, the amount of friction is determined partly by the kind of surfaces touching. When smooth surfaces rub against each other they produce less than do rough surfaces.

If you rub your hands together briskly on a cold day you can feel heat. This heat comes from friction produced by the rubbing of your hands. In some cases friction produces unwanted heat. An electric drill turning on metal can produce enough heat to ruin the tip of the drill. High-speed drills and cutting tools need a cooling system to absorb the heat produced by friction.

It is often necessary to reduce friction in moving objects. Friction reduces the ability of a machine to work at its best and causes parts to wear out, as well.

There are several ways of reducing friction. The more friction that is eliminated in the moving parts of machinery, the longer the machine will last. A rolling object has much less friction than a sliding one (FIG. 2-5). For this reason wheels,

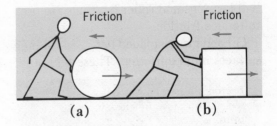

FIGURE 2-5 (a) The figure is rolling a drum. (b) The stick figure is pushing a cart.

BE CURIOUS 2-1:   **Observe the result of forces acting with or against each other.**

Study each of the situations shown in drawings (a) through (i). The big stick figures represent a force of 10 newtons and the small stick figures represent a 5 newton force. In each of the nine cases shown, tell whether the resulting forces are in *equilibrium* or are *unbalanced forces*. If the resulting force is an unbalanced force, what is the *amount* and *direction* of this force?

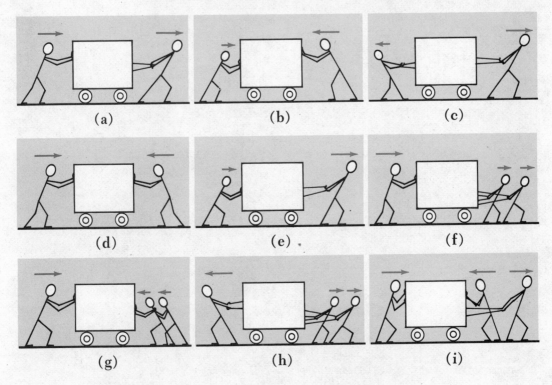

ball bearings and roller bearings are used to reduce friction.

Oil and grease placed between rubbing surfaces reduce friction. These and other substances used for this purpose are called *lubricants*. They allow rubbing surfaces to slide more easily over each other thereby reducing friction and unwanted heat.

A layer of compressed air between surfaces may also act as a friction reducing agent. Air cars and air boats ride on a

> *Bearing* A part on which something rests or in which it turns.

FIGURE 2-6 The Flying Fish, an Italian-built hydrofoil, speeds over the water in a test run. It can carry 60 passengers at speeds up to 40 miles an hour. *(Wide World Photos.)*

cushion of compressed air that eliminates most of the friction, which would otherwise prevent motion (Fig. 2-6).

In some cases friction is necessary and useful. A moving automobile is stopped by friction in the brake linings. It is friction between the tires and a road that enables the automobile to move and stop on the road. You realize this when you try to start or stop a car on an icy surface. You also realize that friction between the sole of your shoes and the surface you are walking on is what enables you to move along when you try to walk on a surface! Why do you slip more easily when you walk on an icy surface?

BE CURIOUS 2-2:   **Measure the force of friction.**

Spring balance
Clay brick
String
Wooden dowels

In this investigation you will use a spring balance to pull a brick across a flat surface. For each part of the investigation observe the force registered on the spring balance while the brick is moving along at a steady speed.

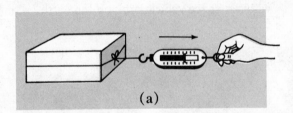

(a)

Part A

Attach the spring balance to the brick, as shown in Figure (a). Pull the brick along a smooth level surface. Try a few practice runs until you can get a steady motion. Then observe and record the force reading on the balance during a smooth run.

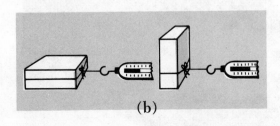

(b)

Part B

Repeat the procedure in Part A turning the brick in the two positions shown in Figure (b). Record the resulting force reading for each. What conclusion can you make from comparing these two forces with each other, and with the force reading in Part A?

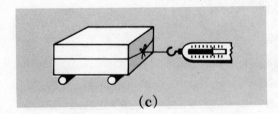

(c)

Part C

Place the brick on two dowels, as shown in Figure (c). Record the force needed to pull the brick along at a steady speed. Explain your results when you compare this force with those in Parts A and B.

A scientist who works in the field of physics is known as a *physicist*. A physicist works with forces in such areas of study as heat, sound, light, electricity, magnetism and mechanics.

A physicist who is concerned with problems in outer space is referred to as an *astrophysicist*. An important concern of astrophysicists is the measurement of the force of gravity in different parts of the universe. They must be able to calculate the forces caused by the hot gases escaping from a rocket engine. By knowing the size and direction of these forces, they can predict the motion of the rocket.

Atomic physicists study the effects of atomic particles as they move around inside the atom. Persons whose studies are con-

centrated on forces within the nucleus of the atom, are called *nuclear physicists*.

Physicists work with other scientists in studying the oceans. They are principally concerned with the forces and motions of water such as tides, currents, and wave motion in ocean waters. This study is called *physical oceanography*.

To study forces inside the earth you must know principles of geology (the study of the earth) and of physics. So, this science is called *geophysics*. A geophysicist studies such things as the forces of earthquakes and what causes them.

The entire universe is the laboratory of the physicist. The research work may go on within a laboratory of some university, or within a space vehicle orbiting the earth.

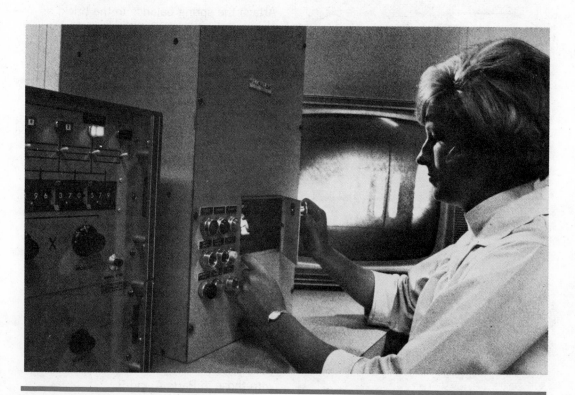

Suppose you are pulling a heavy box to the left across the floor. The force of friction would make it seem as if something is pulling against you to the right (FIG. 2-7a). What if you reversed the direction of your pulling force (FIG. 2-7b)? The force of friction now reverses its direction also. From this observation it is clear that the force of friction is always in the opposite direction to the motion of the object. Thus, for a moving automobile, the force of friction is backward as the car moved forward, and, forward, if the car is backing up.

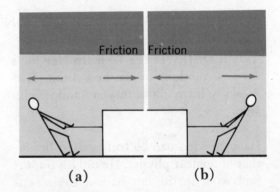

FIGURE 2-7 (a) The stick figure pulls cart in one direction, and (b) in the opposite direction.

OBJECTIVE 2 ACCOMPLISHED? FIND OUT.

1. Define force as a scientific term.
2. What is a newton?
3. When more than one force acts in the same direction, what is the resulting force?
4. What is meant by an object being in equilibrium?
5. What kind of force is needed to set an object in motion?
6. What causes friction?
7. Make a list of ways that friction can be reduced.
8. What is the direction of the force of friction with respect to the direction of the motion of an object?
✸ 9. Write a few paragraphs describing what it would be like in a world where there is no friction.

3 | NEWTON'S LAWS OF MOTION

YOUR OBJECTIVE: To learn Newton's laws of motion; to see how these laws work; to learn about rotation and revolution.

Isaac Newton can be truly called the father of nuclear physics. He did his important work in the last half of the seventeenth century. It was not until our century, the twentieth, that Albert Einstein added some important new ideas to Newton's physics.

Before Newton's time philosophers, who are students of life, wondered about motion. They were concerned about why objects move or why they remain motionless, or at a *state of rest*. Some people thought that a state of rest was the natural state because it seemed for them that some force was necessary to move an object. Other people felt that motion was

| *Law* In science a statement that is always true. |

the natural state because objects tend to drop to the ground. But all agreed that there was a relationship between force and motion.

The true relationship between force and motion was clearly stated for the first time by Newton. This relationship is now called **Newton's first law of motion, which states that an object tends to stay in a state of rest or to move with a constant velocity unless acted upon by an unbalanced force. This property of objects is called inertia. Another way of stating Newton's first law is to say that all objects have inertia.**

Figure 3-1 shows a coin resting on a card placed on top of a glass. When the card is flicked, the coin remains at rest because of its inertia. When the card is rapidly withdrawn, the coin drops into the glass because it is pulled downward by the unbalanced force of gravity.

Newton's second law of motion states that an object accelerates faster as the

FIGURE 3-1

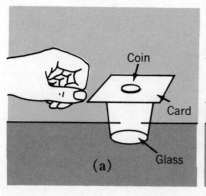

Coin

Card

Glass

(a)

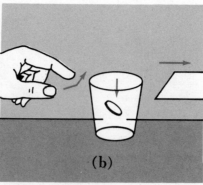

(b)

Sir Isaac Newton was one of the great geniuses of all time. He was born in England on December 25, 1642. By the age of 25 Newton had discovered how gravity worked. A story has been told that Newton discovered the idea behind gravity when he saw an apple fall from a tree. He realized that the force which made the apple fall, also kept the moon in its *orbit*, or circular path, around the earth, He realized that this force of gravity also controlled the motion of planet Earth and the other planets of the solar system in their orbits around the sun.

In the year 1687 Newton's famous book, *Mathematical Principles of Natural Philosophy*, was published. This book, which develops calculus (a branch of mathematics), the idea of gravity and the three laws of motion, was one of the most important books ever written in the field of science.

unbalanced force gets larger, or as the object's mass gets smaller. This can be seen from the equation,

$$\text{acceleration} = \frac{\text{Force}}{\text{mass}} \quad \text{or} \quad a = \frac{F}{m}$$

When you pedal a bicycle, you apply an unbalanced force that results in a forward motion (FIG. 3-2). Since an unbalanced force causes acceleration, the bicycle should accelerate as long as you pedal. This means that the bicycle should move faster and faster without limit. But this is not what really happens! When the bicy-

> *Mass* The quantity of matter contained in an object.

FIGURE 3-2 The stick figure is riding a bicycle. Arrows show the direction of the forces of the pedaling and friction.

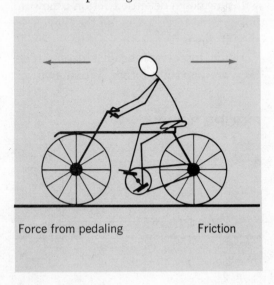

Force from pedaling   Friction

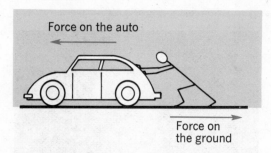

Force on the auto

Force on
the ground

FIGURE 3-3

cle reaches a certain speed, it moves constantly at this speed as long as you pedal at the same speed and with the same force. Can you explain why this is true?

If your answer was "friction," you were correct. The bicycle continues to accelerate until the forward force applied by your pedaling *equals* the backward force caused by friction. (This backward force slows the bicycle down.) When there is no longer an unbalanced force, the bicycle will no longer accelerate.

When you stop pedaling, the bicycle will slow down because friction is now an unbalanced force acting on the bicycle to slow it down.

When forces act against each other, they are opposing forces. When two opposing forces are equal, they are in equilibrium.

If it were possible to remove all friction from the moving parts of a machine, then when the machine is set into motion, there would be no opposing force, so the motion would operate forever. Such a device is called a *perpetual motion machine*. Although many people have tried to build such a machine, no one has ever succeeded.

Suppose you are pushing an automobile. While your hands are applying a forward force on the car, your feet are applying a backward force on the ground (Fig. 3-3). The harder you push the car, the harder your feet push the ground. It would be impossible to apply a forward--pushing force if a slippery surface, like ice, prevented your feet from applying an equal backward force.

When a force is applied in one direction, an equal force is acting in the opposite direction. This is **Newton's third law, which states that for every action, there is an equal and opposite reaction.**

You can move a cart forward by pushing it (Fig. 3-4a). Why would it be impossible to move it if you are standing on the cart while pushing it (Fig. 3-4b)? You

FIGURE 3-4a

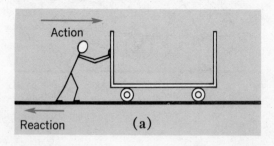

Action

Reaction   (a)

FIGURE 3-4b

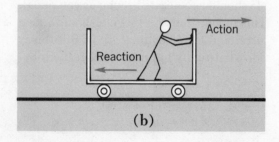

Action

Reaction

(b)

Study the accelerated motion of a pendulum.

String
Watch with
    sweep-second hand
Several different kinds
    and sizes of masses
    that can be used for
    pendulum bobs
Meter stick

Hang a bob (weight) from a string attached to the top of a door frame or another convenient place (see figure). A hanging weight is called a *simple pendulum*. In this investigation you will watch the acceleration of the pendulum. The bob itself is called the *pendulum mass*.

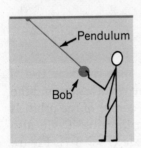

Pendulum

Bob

The stick figure pulls bob up toward the right. The arrow shows the motion of the bob.

Part A

Raise the bob to a measured height above the floor. Hold the string firmly; then let it fall. Watch its swinging motion and try to answer the following questions. What happens to the speed of the bob after you let it go? What force is acting on the bob that causes its motion? When is its speed greatest? When is its speed least? When is it accelerating? How high does it rise above the level of the floor on the opposite side of its swing from the point where it was let go?

Part B

Again let the bob go from a measured height above the floor. Record the time that it takes to make *ten* complete swings. (One complete swing is the motion of the bob from the starting point to the opposite side and back again to the starting point.) How much time does *one* swing back-and-forth take?

Now replace this bob with a larger one, keeping the same length of string. Record the time that the new bob will take to make one complete swing. Now let it go from the same height above the floor, as before, and check the time for one swing. Can you explain what happened?

Part C

In Part B you changed one of the *variables* (a changeable condition, such as temperature) of the pendulum, the mass. What are some of the other variables that can be changed? Now change one of the other variables and note any change in the motion. (You can change only *one* variable at a time, otherwise you will not be able to tell which one caused the change in motion.) How do some of the variables affect the motion of the bobs?

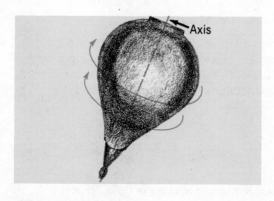

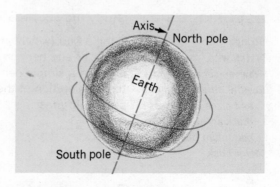

FIGURE 3-5

FIGURE 3-6

are applying *two* forces on the cart. The *action* comes from your hands pushing it forward, and the *reaction* comes from your feet pushing it backward at the same time. Since action and reaction are opposing forces, these forces are balanced, therefore the cart is in equilibrium so it will not move. Remember that an unbalanced force is needed to accelerate the cart from zero speed to a higher forward speed.

The motion of a spinning object, like a top, is called rotation. An object in rotation is said to *rotate*. **A rotating object moves about an imaginary line through its center. This line is called an axis** ('ak-səs). The top in Figure 3-5 is rotating, or spinning, about its axis. The earth, moon, planets and sun each rotates about its own axis. The Earth's axis passes through the north and south poles (FIG. 3-6).

Another kind of circular motion is called revolution. An object in revolution is said to *revolve*. A revolving object moves along a circular path called an **orbit.** You could think of the orbit as the tire on a bicycle wheel. Then when the

wheel moves, the tire revolves around the axle, which is at the center of the orbit. If there is a pebble on the tire, it will also revolve about the axle (FIG. 3-7). Figure 3-8 shows the moon in its orbit around the Earth. **A force that pushes along a circular path or causes spinning motion is called a torque** ('tȯ(ə)rk).

A force that pulls a revolving object toward the center of its orbit is called a centripetal force. The word *centripetal* has two main syllables — *centri*, meaning "center"; *petal* comes from a Latin word meaning "to seek," just as in the word *peti*tion. Therefore *centripetal* means "seeking a center." The Earth's gravity is an example of a centripetal force acting on the moon because it keeps the moon in orbit around the Earth.

You might now think that there is an opposing force which would tend to pull a revolving object *away* from the center. Such a force could be called *centrifugal*, or "fleeing from the center"; *fugal* comes from a Latin word meaning "to flee", just as in the word *fugitive*. You would be aware of such a force if you are riding in a

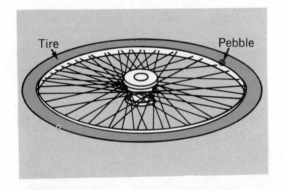

FIGURE 3-7

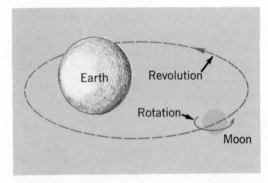

FIGURE 3-8

vehicle that suddenly turns to the right and you seem to be "pulled" in the *opposite* direction, to the left. However, what is actually happening is that you are trying to remain on a straight-line course while this vehicle is turning. It *seems* that you are pulled in the opposite direction. This is a good example of *inertia*. **There is really no centrifugal force, but only an illusion of force due to inertia.**

OBJECTIVE 3
ACCOMPLISHED?
FIND OUT.

1. Newton's first law is sometimes called the law of inertia. What is inertia?
2. State Newton's second law. (You may use the equation form.)
3. State Newton's third law.
4. You are riding a bicycle along a level surface. Why do you have to keep pedaling to keep the bicycle moving at constant speed?
5. Describe the motion of a rotating object.
6. Describe the motion of a revolving object.
7. What is torque?
* 8. Describe centripetal force. What causes "centrifugal force?"
* 9. An unbalanced force causes an object to accelerate. What is the acceleration of this object if the force causing acceleration is doubled?
* 10. What is the force needed to accelerate a mass of 3 kilograms at 6 meters per second per second (6 m/sec/sec)?

4 | WHAT IS MOMENTUM?

YOUR OBJECTIVE: To find out what momentum is; to understand what happens to momentum in explosions and collisions.

Which of these objects would you *least* like to catch: a lightweight object moving rapidly, a heavy object moving slowly, or a heavy object moving rapidly? Naturally your choice would be the last one. This object would be the hardest to catch because of its heavier mass and rapid velocity that give the object *momentum*.

Momentum is the product of mass and velocity; that is, momentum $(P) = mass \times velocity$. (The letter p will be used for momentum since m is used for mass.) This equation can be shortened to

$$p = mv$$

where p is the momentum, m is the mass, and v is the velocity.

You may see a similarity between this equation and Newton's second law described in Section 3. Recall this law $a = F/m$, or $F = ma$. You see here that the force F is the product of mass m and acceleration a. Hence force *depends* on mass and acceleration, whereas momentum depends on mass and velocity.

You have seen that mass is measured in kilograms and velocity is measured in meters per second. Since momentum is a product of these units, then momentum is measured in kilogram-meters per second. This is abbreviated as kg-m/sec.

Ordinarily you think of an explosion as a sudden loud blast followed by vibrations, such as might result when extreme heat is applied to dynamite. Probably you have often seen the small explosions produced by firecrackers. **An explosion happens when objects are forced away from each other by some type of energy** (FIG. 4-1). Some other examples of explosions are: a runner leaving a starting line, a mouse trap snapping shut, or a person jumping off a diving board.

SAMPLE PROBLEM: An object with a mass of 5 kg is moving with a velocity of 6 m/sec. What is its momentum?

Solution: Replace the m for mass with 5 kg and the v for velocity with 6 m/sec in the momentum equation

$$p = mv$$

Then,

$$p = 5 \text{ kg} \times 6 \text{ m/sec.} = 30 \text{ kg·m/sec.}$$

FIGURE 4-1 This nuclear explosion was part of a test series made in July, 1946, at Bikini Atoll. *(Official U.S. Air Force Photo.)*

In Figure 4-2, a coil spring is squeezed together and placed between two toy carts that are tied together with a string. If the string is cut, the spring will quickly open and push the carts apart. This explosion, in turn, gives *momentum* to each cart. The spring will push both carts with the *same* amount of force.

Newton's third law of motion tells us that the spring causes the same amount of force on both carts. If the force on one cart is the action, then the force on the other cart is the reaction. And both forces are equal, but opposite in direction.

FIGURE 4-2

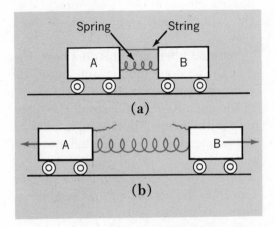

FIGURE 4-3

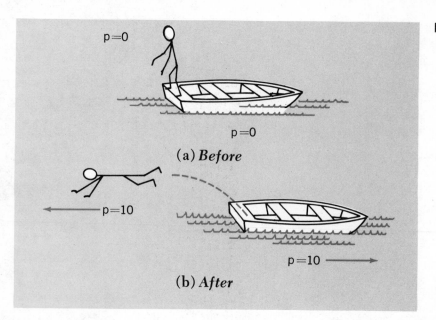

p=0

p=0

(a) *Before*

p=10

p=10

(b) *After*

If the carts in Figure 4-2 have the *same* mass, after the explosion their velocities — the direction of motion — will also be the same. Can you tell what would happen if cart A has a greater mass than cart B? The correct answer is that the larger mass would have a slower velocity. Therefore the larger mass will move slower than the smaller mass after the explosion. However the *momentum* of each cart will be equal and opposite in direction.

Suppose a person jumps off a boat (FIG. 4-3). By Newton's third law of motion, the action of jumping causes the boat to move in the opposite direction from the direction of the jump. This is the reaction. If the jumper has a certain momentum resulting from the jumping action, the boat will receive the same amount of momentum from the action, but in the opposite direction.

Fuel burning in a rocket engine is an-other form of explosion. The explosive force gives the resulting hot gases a downward velocity. The rocket then moves with an upward velocity (FIG. 4-4). Since the rocket has the larger mass, it will have a slower velocity than the escaping gas moving in the opposite direction. Again the *momentum* of the rocket and the momentum of the gas are equal and opposite in direction.

There are two types of collision, *elastic* and *inelastic*. A rubber ball bouncing from a hard surface, such as a floor or wall, is an example of an elastic collision. A lump of clay that falls to the floor and sticks there is an example of an inelastic collision. **In both (elastic and inelastic collisions, one object (mass) may give momentum to the other object. The amount of momentum given depends on the masses of the colliding objects and on whether the collision is elastic or inelastic.)**

FIGURE 4-4 The first Saturn space vehicle was launched in a flight that lasted about 8 minutes, 3.6 seconds. It reached a velocity of 3, 607 miles per hour and a height of about 84.3 miles. The distance traveled was 214.7 statute miles. (*NASA.*)

Find out what happens to the momentum of objects when they collide.

2 glass marbles (same size)
2 metal marbles (same size)
Grooved plastic ruler

In this investigation you will cause collisions between solid metal and glass marbles. Watch changes in the motion of these marbles after the collision.

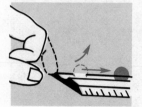

Part A

Place one of the glass marbles in the groove of the plastic ruler. With your finger snap a second glass marble of the same size to make it move along the groove and collide with the marble at rest (see the figure). Describe the motion of the two marbles after the collision. Now use the two metal marbles of the same size and snap one against the other. What can you say about collisions between equal masses?

Part B

Repeat the procedure used in Part A. But this time make the glass marble collide with a metal marble at rest. Describe the motion of the marbles after the collision.

Part C

Again repeat Part A. However this time make the metal marble collide with the glass marble at rest. Describe the motion of the marbles after the collision as you did in Part B. Compare your result here with that in Part B.

In Be Curious 4-1 you found that the collisions of the glass and metal marbles were elastic. After the collisions, one marble always bounced off the other. At no time did one stick to the other. How they bounced depended on which of the two masses was larger. In the case of equal masses, one marble gave all of its momentum to the other. This also happens when pool balls strike head on.

Consider what happens when a marble with a larger mass strikes one with a smaller mass. Only part of its momentum is given to the smaller mass. The metal marble continued rolling in the same direction after the collision, but with less speed than before. This is also the type of collision that occurs when a bowling ball strikes lighter pins. What happens to the bowling ball?

Throw a tennis ball against the back of a parked truck. The ball will bounce back. And the truck will move forward very slightly. This is an example of a very small mass colliding with a very large one.

Consider what happens when a marble with a smaller mass is struck by one with a larger mass. The marble with the smaller mass will keep some momentum, but it will be in the opposite direction, as before.

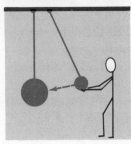

OBJECTIVE 4 ACCOMPLISHED? FIND OUT.

1. Define momentum.
2. Two carts are held together by a string (FIG. 4-2). When the string is cut, the spring will push the carts apart. If the carts have the same mass, how would their velocities compare as they move away from each other? How would the velocities compare if cart A has a larger mass than cart B?
3. Two metal globes of different masses are hanging from the ceiling (see the figure). Lift the smaller globe, and then let it go so that it swings down and hits the larger globe. What will happen after the globes collide? What will happen if the larger globe is lifted and let go?
4. Suppose it is winter and you are standing on some ice in the middle of a pond. Also suppose the ice has a perfectly frictionless surface. This would make it impossible for you to walk in any direction. How could you get to the edge of the pond?
5. One object has a mass of 4 kg and a velocity of 2 m/sec. Another object has a mass of 2 kg and a velocity of 4 m/sec. Compare the momentums of both objects.
✻ 6. Suppose cart A in FIG. 4-2 has a mass of 2 kg and cart B has a mass of 4 kg. When the spring pushes the carts apart, cart A moves to the left with a momentum of 12 kg-m/sec. What is cart A's velocity? What is the momentum and direction of cart B? What is cart B's velocity?

5 | NEWTON'S LAW OF GRAVITY

YOUR OBJECTIVE: To learn how Newton's law of gravity explains weight; to study the acceleration of a falling object caused by gravity.

It has been known for thousands of years that objects tend to fall to the earth. When you stumble, you tend to fall to the ground. An object always falls toward the center of the Earth no matter where the falling motion is taking place.

You have already learned that motion in any direction is an action which is caused by some unbalanced force. You may guess that a falling motion is no exception. It too must be the result of an unbalanced force, but acting in the direction toward the Earth's center. This force is called *gravitational force*, or simply *gravity*.

After Newton had watched many objects falling and the motion of heavenly bodies, he was able to state his **universal law of gravitation: All objects are attracted to each other by a force that is directly proportional to the masses of the objects, and inversely proportional to the square of the distance between them.** This is more fully stated as follows:

1. As the mass of either object becomes larger, the force attracting them also becomes larger. Conversely, as the mass of either object becomes smaller, the force attracting them also becomes smaller.
2. As the distance (squared) between two objects becomes larger, the force attracting them becomes smaller. Conversely, as the distance (squared) between the two objects becomes smaller, the force attracting them becomes larger.

Newton pointed out that the force of gravity acts on objects near the Earth and on distant objects, such as planets, the moon and stars. However, you will now be concerned only with this force on objects on or near the earth's suface.

If you hold a tennis ball about 3 ft. from the ground, you may be aware of a small force which tends to pull the tennis ball toward the ground. If you let the ball go, this force is now unbalanced and it will fall to the ground. No matter how often or where you repeat this action, the ball will drop to the ground. You will find that its direction of motion is always toward the Earth's center.

How do you weigh an object? When a clerk at the meat counter of a supermarket weighs a cut of meat, it is usually placed on a spring scale. The meat's weight causes a spring to move a pointer on the scale that reads in pounds and ounces.

As you may have already guessed, weight and gravity are related. Weight is the measurement of the gravitational force which pulls an object toward the earth's center. If the meat on the scale weighs 2¼ pounds, this is the same as saying that the meat is pulled toward the Earth with a force of 2¼ pounds.

After you have taken the cut of meat weighing 2¼ pounds from the store,

BE CURIOUS 5-1: Find out if heavy objects fall faster than light ones.

2 sheets of paper
Rubber ball

Part A

In this part of your investigation compare the falling speed of objects with the *same* mass. Drop two sheets of paper from a height of about 2 ft. Which one reaches the floor first? Fold one of the sheets in half and leave the other sheet unfolded. Which sheet falls faster? Now fold the folded sheet in half again so that it has three folds and repeat the test. Compare your three results and explain why they are different.

Part B

In this part of the investigation compare the falling speed of objects with *different* masses. This time drop a sheet of paper and a rubber ball. Which has the larger mass? Drop the sheet of paper and the ball at the same time. Which fell faster? Now crumple the sheet of paper into a tight ball. Drop the paper ball and the rubber ball at the same time. Which fell at a faster rate? Can you explain the results?

where it was weighed, and brought it home, the weight of the meat will not have changed unless there has been a large change in the distance between the meat and the Earth's center. This might happen if your home is on a mountain and the store at the valley. Then a difference between the meat's weight at the store and its weight at home might be seen if a very *fine* scale is used. The weight and mass of an object are not really the same thing. The mass of the meat does not change no matter where you take it on or off the Earth's surface. But the weight of the meat will change if the distance between it and the Earth's center has changed. **As the distance above the Earth's surface becomes greater, the weight of an object becomes smaller. As this distance becomes smaller, the weight becomes greater.**

The force of gravity causes objects to fall to the earth. When gravity provides an unbalanced force, an object acted upon accelerates. If you drop a very light object, such as a feather, and a glass marble, at the same time, which do you think will reach the floor first (Fig. 5-1)? The

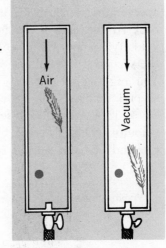

FIGURE 5-1

results of your investigation in *Be Curious 5-1* should tell you that the marble will reach the floor first. This is because air greatly slows down the fall of the feather. If you dropped the feather and the marble in a place with no air, you would see that they fall at exactly the same speed. **An object dropped in a vacuum accelerates at 32 ft/sec/sec or 9.8 m/sec/sec.** A swimmer falling from a diving board, a hammer falling to the ground and a book falling to the floor, all accelerate at nearly 32 ft/sec/sec in air. Objects like a sheet of paper or a feather fall at a much slower rate because the air slows down their fall. However, generally, air does not slow down the fall of objects with large masses very much.

An object dropped in a vacuum on the Earth's surface accelerates at 32 ft/sec/sec. It continues to accelerate at this rate until it hits the ground. TABLE 5-1 shows an object's velocity at each second between 0 and 10 seconds of its fall. You can see (in the second column from the left) that the velocity increases by 32 ft/sec from one second to the next.

A skydiver jumping from an airplane accelerates at nearly 32 ft/sec/sec (FIG. 5-2). At the end of the first second the velocity is about 32 ft/sec. At the end of the second it is 64 about 32 + 32 ft/sec. You might expect, at the end of the third second, that the velocity would be about 32 + 32 + 32 = 96 ft/sec and that the velocity would increase by 32 ft/sec at the end of each second until the para-

TABLE 5-1	
An object falling in a vacuum on the Earth's surface.	
Time (sec)	Velocity (ft/sec)
0	0
1	32
2	64
3	96
4	128
5	160
6	192
7	224
8	256
9	288
10	320

chute opens. However this would *only* be true if the jump were made in a *vacuum*.

Because the skydiver is falling through air, there will be two forces acting on the diver's body. One is *gravity*, which accelerates the fall, the other is friction caused by air, which slows down the fall. As the fall becomes faster, the force of the air opposing the fall becomes greater. However the force of gravity remains the same throughout the fall. When the force of the air equals the force of gravity the two opposing forces are balanced. At this point the skydiver continues to fall at a *constant* or unchanging velocity, since balanced forces do not cause acceleration.

Vacuum A completely empty space.

FIGURE 5-2 Skydiver in free fall after jumping from an airplane. *(Parachutes, Inc.)*

Measure the acceleration due to gravity.

Pendulum bob
String
Timing clock

A device or instrument used to measure the acceleration caused by gravity is called *gravimeter* (gra-ˈvim-ət-ər). A simple pendulum may be used as a gravimeter because its swinging motion is caused by gravity. Physicists have studied the motion of pendulums and have found that the period (time it takes the pendulum to make one complete swing) depends on only two things: These are the *length* of the pendulum, and the gravitational force acting on it.

This relationship is expressed by the equation $T = 2\pi\sqrt{1/g}$. In this equation T is the time a pendulum takes to make one complete swing, l is the length of the pendulum, and g is the acceleration due to gravity for the place where the pendulum is swinging.

The acceleration of gravity can now be measured by measuring the length of the pendulum you have set up, and the time T it took for one complete swing. Substituting these two values in the equation and solving for g will give you the correct value for the acceleration.

The equation $T = 2\pi\sqrt{1/g}$ can be rewritten as

$$g = 4\pi^2 \times l/T^2.$$

Since $4\pi^2 = 39.4$ then

$$g = \frac{39.4 \times 1}{T^2}$$

The units for the acceleration depend upon the units used to measure the length l and the time T. If you measure the length in meters and the time in seconds, then the acceleration will be in m/sec/sec.

OBJECTIVE 5
ACCOMPLISHED?
FIND OUT.

1. What is Newton's universal law of gravitation?
2. What is weight? How does it differ from mass?
3. Why does a large rock weigh more than a small one?
✿ 4. Why would you weigh less on the moon than on the Earth?
5. The acceleration of an object due to gravity at the Earth's surface is 32 ft/sec/sec. How fast will a falling object move after the first second? The second second? The third second (assume there is no air)?
6. The Apollo spacecraft used a parachute to slow its fall to the Earth's surface. Why was it impossible to use a parachute for the moon landing?

6 | PROJECTILES AND SATELLITES

YOUR OBJECTIVE: To learn about projectile and satellite motion; to understand what weightlessness means.

If you drop a stone from the edge of a cliff, it will fall downward in a straight-line motion. If you throw a stone, it will follow a curved path. The amount of force you use in throwing and the angle at which it is thrown will determine the path the stone will follow. A stone or any other object that is thrown is called a *projectile.*

The motion of a projectile thrown in a *horizontal* direction is shown in Figure 6-1 moving toward the right. A similar object is dropped at the *same* time at the left. Both objects are falling at the *same* speed because of gravity. So they will both strike the ground at the same time.

The same thing is true for a bullet fired in a horizontal direction. If you could drop a bullet straight to the ground, and at the same time and from the same height you fired another bullet from a rifle, you would find that both bullets reach the ground at exactly the same time. If the rifle bullet is pointed upward from the horizontal position, the bullet will reach the ground later than the bullet dropped from the same height.

Newton thought of an interesting type of projectile motion. He imagined a cannon ball being fired in a horizontal direction from a mountain top. He thought

Horizontal Level with the earth's surface.

that if the cannon ball could be made to go faster and faster it would keep going farther before reaching the ground, and after a while the speed would be great enough to put the ball in orbit around the earth. And, if the path of this orbit were far enough above the Earth, so that no air friction could slow down its motion, the cannon ball would continue in this orbit. Hence it would become an Earth satellite.

Since Newton's time, scientists have tried to find out how a satellite could be put into orbit around the Earth. The main problem was to get rid of air friction which would slow down the projectile as it moved through the air. They thought

FIGURE 6-1

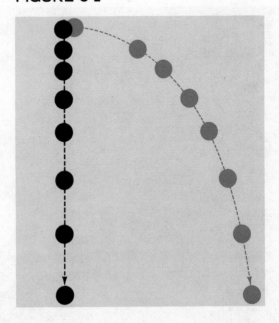

that if a projectile could be lifted high enough, it could pass above the air layer around the Earth. Then a horizontal force could give the projectile enough speed for it to continue in orbit around the earth. For this to be possible it would first be necessary to build a rocket that could be fired to this great height.

In 1957, Soviet scientists fired a rocket far enough to pass through the Earth's air layer. The rocket then fired a projectile weighing 184 pounds with a horizontal speed of about 5 miles per second (18,000 mi/hr). This projectile, which became an earth satellite — like the moon — was called "Sputnik." **The term satellite refers to an object that orbits the Earth or any other planet or star.** This was the first artificial satellite. Since then many different satellites have been sent into orbit in space by both the United States and the Soviet Union.

If there were no forces acting on an Earth satellite, it would move at a constant speed in a straight-line motion away from the Earth. The force that keeps a satellite in orbit around the Earth is gravity. This force always pulls a satellite toward the Earth's center. The *inertia* of the satellite, which tends to keep it moving in a straight-line path away from the Earth, is balanced by gravity. Therefore, a satellite is kept in orbit (Fig. 6-2). If a satellite's speed is too great, it will escape from the Earth. If the speed is too low, it will fall to Earth.

FIGURE 6-2 Overhead view of the Skylab space station contrasted with the black sky in the background. *(NASA.)*

FIGURE 6-3 (a) The stick figure is in a standing elevator — gravity pulls the ball to the floor and the scale shows stick figure's correct weight. (b) The stick figure is in a freely falling elevator — gravity is pulling the elevator as well as the stick figure and the ball, so the stick figure and the ball seem to be weightless, whereas they are also falling with the elevator — at the same speed as the elevator.

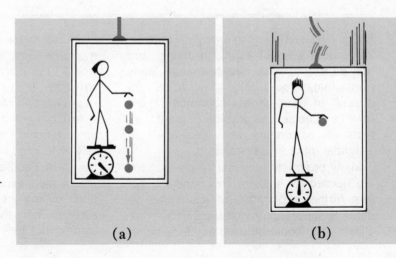

(a) (b)

Suppose you are asked to perform two experiments in an elevator: (1) weigh yourself on a scale, and (2) drop a ball (FIG. 6-3). First these experiments are done while the elevator is standing still. The same experiments are then done again, but when the elevator is falling freely, that is, when the elevator is accelerating downward at 32 feet per second. For this to happen, the elevator cable must be cut and the elevator is falling with no air friction to slow down its fall.

What would be the results of these experiments when the elevator is standing still? You would be correct if you said that your weight would be the same as it would be on any scale on the Earth's surface. The ball would also drop and strike the elevator's floor just as it does when you are standing any place on the Earth.

However, when the elevator is falling, it would not be the same story (see FIG. 6-3b). You would find that the scale reads zero, and that the ball would not leave your hand when you let it go — it would

hang in the air. It would seem that there is no gravity. But what is really happening?

Although gravity has not left the falling elevator, you are experiencing *weightlessness*. There seems to be a force that causes equilibrium and cancels gravity. The acceleration seems to be zero inside the falling elevator even though the ball is falling freely, but so are you and the elevator, as well.

An astronaut inside an orbiting satellite also experiences weightlessness for the same reason. Like the falling elevator, the satellite and the astronaut inside of it are actually in free fall.

To the astronaut in the orbiting satellite, there seems to be no gravity. A scale would indicate that his weight is zero. Objects dropped inside the satellite do not fall. But there is a gravitational force acting all the same. If not, the satellite would not stay in orbit, but travel in a straight line away from the Earth. Objects could have weight if a satellite rotates.

Imagine yourself living in a weightless state. Your usual ways of eating, drinking, and moving around would have to change. Many of the things you usually do would be very difficult in a weightless condition. There have been no serious physical or mental effects on astronauts who have lived in weightlessness for several days.

Many scientists feel that *space stations* are needed for studying this condition before further exploration of space can be done. Space stations orbit the Earth as satellites. They could be used as stopping-off places for space vehicles on their way to the moon or to other planets. These stations would have many useful things found in a small city, because people in space stations might have to spend a long time there.

Artificial gravity could be given to a space station, if the station could be made to rotate on its own axis while it is in orbit around the Earth. People inside the spinning station would then experience a "centrifugal action" pulling them toward the floor of the station. Then they could do many of the same duties they did on the Earth's surface.

In preparation for space flights astronauts underwent many tests to determine the effects weightlessness would have on them. In this weightlessness test, one airman prepares to walk on the cabin ceiling of an Air Force transport plane, while the other floats through the air holding a gyroscope.

1. Rifle A and rifle B are fired at the same time. Both rifles are fired in a horizontal direction over a level surface. The bullet leaving rifle A has twice the speed as that of the bullet from rifle B. Which bullet will strike the ground first? Which bullet will travel the greater distance?

2. What will happen to an object if its speed is too great to be put into orbit around the Earth by gravity? What if its speed is too low?

3. Explain why the person in the elevator in Figure 6-3b does not see the ball falling.

4. Why does an astronaut in a satellite feel weightlessness?

✿ 5. Write a few paragraphs on what you think it would be like to live in a condition of weightlessness. What problems would you likely meet?

7 | FIELDS OF FORCE

YOUR OBJECTIVE: To learn about force fields and how they apply to gravity, electric charges and magnetism.

A force field is a region where forces act on objects. It is possible to show the size and direction of these forces. Forces are often represented by arrows when a drawing is used to identify them. The arrow heads point in the direction which a force acts.

The gravitational field surrounding the Earth is nearly the same for all places on the Earth. The Earth's gravitational field is represented by arrows in Figure 7-1. The arrows all point in the same direction, toward the Earth's center. This tells

FIGURE 7-1

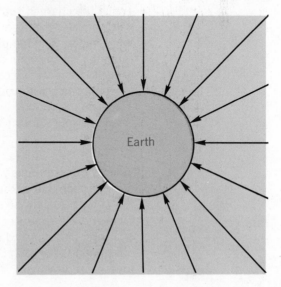

you that any object placed in this field will be pulled toward the Earth — in the direction of the arrows. Notice that the arrows get closer together as they get closer to the Earth. This indicates that the gravitational field is stronger near the Earth.

Pass a pocket comb through your hair. If your hair is dry, and the air is not too damp, the comb will attract small bits of paper. You can get the same results by rubbing a plastic rod on a wool shirt or sweater. The force attracting the paper is called an electric force.

BE CURIOUS 7-1: **Observe the behavior of electric forces.**

Pith balls
Glass rod
Rubber rod
Silk cloth
Wool cloth

In this investigation observe the effects of one type of charge on another. Charge suspended pith balls using a glass or rubber rod.

Part A

Take away any charge on the suspended pith balls touching them with your finger to make them electrically neutral. Rub the rubber rod with wool cloth. What type of charge is now on the rod? Bring this charged rod close to, but not touching, each of the pith balls. What happens when the charged rod is near (but not touching) either pith ball? Repeat these steps but rub the glass rod with the silk cloth. What happens when this charged rod is brought near, but not touching, either pith ball?

Part B

Touch the charged rubber rod to both pith balls. This gives each ball a charge of the same kind as the charge on the rubber rod. What happens to the pith balls? Charge the rubber rod again by rubbing it in the wool cloth, and bring the rod near each pith ball, in turn. What do you observe? Take the charge off each pith ball by touching them with your finger. Now repeat the procedure using the glass rod rubbed in silk cloth. Write down what you observe. What are your conclusions resulting from your observations?

Part C

Touch both pith balls with your fingers to make them neutral once again. Touch one of the balls with the charged rubber rod. Touch the other ball with the charged glass rod. What happens to the pith balls in each case? Recharge each rod and bring each, in turn, near one pith ball and then the other. Write down your observations for each case. What conclusions can you make from these observations?

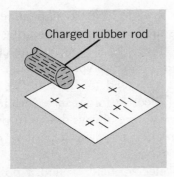

Charged rubber rod

FIGURE 7-2

When certain kinds of materials are rubbed against each other, *electrons,* which are electrically charged particles, may leave one object and be picked up by another object. Both objects are electrically charged. The object that loses electrons will have a *positive* charge. This may be shown by a plus sign +. The object that picks up electrons will have a *negative* charge, which may be shown by a minus sign —.

A rubber rod rubbed in wool or fur will pick up electrons. Therefore the rubber rod will have a negative charge. On the other hand, a glass rod rubbed in silk will lose electrons. Thus it will have a positive charge.

Suppose a negative charge on a rubber rod is touched to a small object that carries no charge, or is *electrically neutral.* That object will receive some of the electrons from the rubber rod, and becomes negatively charged. If a positive charge on a glass rod is touched to the small electrically neutral object, the object will lose some of its electrons and become positively charged (FIG. 7-2).

A charged object can be made neutral by removing the charge on it. If the charge is *positive,* electrons must be added to it. If the charge is *negative,* electrons must be removed from it.

The charge on an object may be removed by touching it with your finger or some large neutral object (FIG. 7-3). An-

FIGURE 7-3 A spectacular example of this is lightning. In a thundercloud the warm air and cold air and water droplets are in violent motion. This causes friction, which "pulls off" electrons. When a large electrical charge builds up, it discharges the stream of electrons. The negatively charged electrons are attracted to a positively charged cloud or neutral earth. Sometimes the lightning discharge is between one cloud and another. Sometimes it is between a cloud and the ground. Is the lightning discharge above between cloud and cloud or cloud and ground? *(U.S.D.A.)*

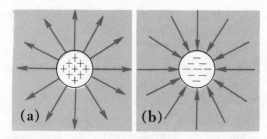

FIGURE 7-4

other way is to lead electrons to or away from the charged object by some conductor (copper wire) attached to the ground by way of a water pipe.

You found in *Be Curious 7-1* that **like electric charges repel each other.** A positive charge repelled another positive charge. A negative charge repelled another negative charge. You also found that **there is a force of attraction between unlike charges.** A positive charge attracted a negative one. A negative charge attracted a positive one.

The *electric field* around an electric charge may have arrows pointing either way — toward or away from the charged body. In order to avoid confusion, scientists have agreed to draw arrows representing electric forces in a certain way. The arrows are always drawn to point in the direction in which a positive charge would move.

The electric field around a positively charged object is shown in Figure 7-4a. The arrows point *away* from the object. Notice that the arrows are also closer to each other near the surface of the object. This indicates that the electric field is *stronger* there.

The field around a negatively charged object is shown in Figure 7-4b. The arrows point *toward* the object. In what direction would a positive charge move in this field?

Over 2,000 years ago, it was discovered that a rocklike material found in nature had the unique property of attracting iron

BE CURIOUS 7-2: **Observe the behavior of magnetic forces.**

2 bar magnets string

Part A

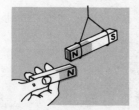

Suspend a bar magnet as shown in the figure. Bring the *north* pole of a second bar magnet near the *north* pole of the suspended magnet. What happens? Bring the *south* pole of the second magnet near the *south* pole of the suspended magnet. What happens now? What can you say about the force between like magnetic poles?

Part B

Bring the *south* pole of the magnet near the *north* pole of the suspended magnet. What happens? What can you say about the force between unlike magnetic poles?

to itself. The material was called *magnetite*. The force of attraction for iron in magnetite was referred to as *magnetism*.

Scientists have since discovered how to treat iron to make it have the same property of magnetism as magnetite ore. It was simply done by stroking an iron bar or rod with magnetite. Much later it was found that stronger magnets could be made by using electric currents.

If a bar magnet is suspended, it will point in an almost north-south direction. Furthermore, the same end of the magnet always points toward the north. This end of the magnet is the *north-seeking pole*, or simply the *north pole*. Accordingly, the opposite end of the bar magnet is called the *south-seeking pole*, or *south pole*.

Like magnetic poles repel each other with a similar kind of force that is between like electric charges. Unlike magnetic poles attract each other with a similar kind of force that is between unlike electric charges. For this reason, the

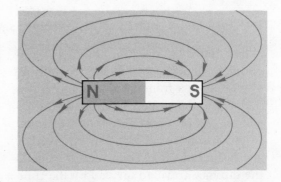

FIGURE 7-5

arrows representing a magnetic force field can also be drawn in either direction. Scientists have agreed to draw the arrows of a magnetic field so that they always point in the direction a *north pole* would move when placed in that field. The magnetic field around a bar magnet is shown in Figure 7-5. Note especially the part of the field near the south pole. Compare this area to the electric field around a negatively charged object (Fig. 7-4b) and the Earth's *gravitational field.*

OBJECTIVE 7
ACCOMPLISHED?
FIND OUT.

1. What is a force field?
2. Why do the arrows on the gravitational field shown in Figure 7-1 point toward the earth?
3. What causes the object in *Be Curious 7-1* to receive an electric charge?
4. The arrows in Figure 7-4b are closer to each other in that part of the field nearer the negatively charged object. What does this tell you about the electric force field?
5. Fill in the blanks in the following incomplete statements:
   a. A positive charge will _____ another positive charge and _____ a negative charge.
   b. A magnetic north pole will _____ another north pole and _____ a south pole.

IN THIS UNIT YOU FOUND OUT

When the motion of an object is compared to something else, that something else is called a frame of reference.

Speed is defined as the distance an object travels in a certain amount of time. Instantaneous speed is the speed given at any instant of time. The average speed is the distance traveled divided by the time it takes to make the trip.

To a scientist the word velocity means not only how fast something is moving but also in what direction the motion is taking place. Velocity includes both speed and direction.

The scientist defines acceleration as a change in speed or velocity occuring during a certain amount of time.

Sometimes an object appears to be displaced when it is really standing still. This false appearance of motion is called parallax.

Force is defined as a push or pull applied to an object. Physics is the science that deals with forces and how they affect matter. When forces act in the same direction, the resulting force is equal to the sum of the individual forces. When forces oppose each other, a force pushing to the right will cancel part of a force pushing to the left. When two equal forces oppose each other, one force cancels the other force. An object at rest needs an unbalanced force to set it in motion. If an object is in motion with a constant speed or velocity, it also will need an unbalanced force acting upon it to change its speed or velocity. Therefore, an unbalanced force is needed to change the motion of an object.

The force of friction causes moving objects to slow down. It also makes it more difficult to keep moving objects going at a constant speed. Since friction is caused by surfaces rubbing against each other, the amount of friction is determined partly by the kind of surfaces touching.

Newton's first law of motion states that an object tends to stay in a state of rest or to move with a constant velocity unless acted upon by an unbalanced force. This property of objects is called inertia. Another way of stating Newton's first law is to say that all objects have inertia.

Newton's second law of motion states that an object accelerates faster as the unbalanced force gets larger, or as the object's mass gets smaller. When forces act against each other, they are opposing forces. When two (or more) opposing forces are equal, they are in equilibrium.

When a force is applied in one direction, an equal force is acting in the opposite direction. Newton's third law states that for every action, there is an equal and opposite reaction.

The motion of a spinning object is called rotation. A rotating object moves about an imaginary line through its center which is called an axis.

Another kind of circular motion is called revolution. A revolving object moves along a circular path. A path of revolution is called an orbit. A force that pushes along a circular path or causes spinning motion is called torque.

A force that pulls a revolving object toward the center of its orbit is called a centripetal force. There is really no centrifugal force, but only an illusion of force due to inertia.

Momentum is the product of mass and velocity. An explosion happens when objects are forced away from each other by some type of energy.

In both elastic and inelastic collisions, one object (mass) may give momentum to the other object. The amount of momentum given depends on the masses of the colliding objects and on whether the collision is elastic or inelastic.

Conservation of momentum occurs in all collisions or explosions. This means that when objects (masses) collide, the total momentum of the objects is the same after collision as it was before the collision had taken place. The direction of the momentum is the same as the direction of the velocity of a moving object. Either type of collision gives momentum to both objects. When objects have equal masses, each object will give all of its momentum to the other object.

Newton's universal law of gravity states that all objects are attracted to each other by a force that is directly proportional to the mass of the objects and inversely proportional to the square of the distance between them. Therefore, as the distance above the Earth's surface becomes greater the weight of an object becomes smaller. As this distance becomes smaller, the weight becomes greater.

The force of gravity causes objects to fall to the Earth. When gravity provides an unbalanced force, an object acted upon accelerates. An object dropped in a vacuum accelerates at 32 ft/sec/sec or 9.8 m/sec/sec.

The term *satellite* refers to an object that orbits the Earth or any other planet or star. If there were no forces acting on an Earth satellite, it would move at a constant speed in a straight-line motion away from the Earth. The force that keeps a satellite in orbit around the Earth is gravity. This force always pulls a satellite toward the Earth's center.

A force field is a region where forces act on objects. Like electric charges repel each other. There is a force of attraction between unlike charges. Like magnetic poles repel each other with a similar kind of force that is between like electric charges. Unlike magnetic poles attract each other with a similar kind of force that is between unlike electric charges.

UNIT OBJECTIVES ACCOMPLISHED? FIND OUT.

Part A Match the numbered phrases in the left-hand column with the lettered terms on the right.

1. The distance traveled by an object in a certain amount of time

a. Vacuum
b. Velocity
c. Force

2. Any push or pull
3. The tendency of an
   object to stay moving
   at a constant
   velocity
4. The property found by
   multiplying an object's mass
   by its velocity
5. Spinning on an axis
6. A property that has
   both speed and
   direction
7. A measurement of the force
   of gravity
8. An object held
   in orbit because
   of gravity
9. Two (or more) objects are
   forced apart by energy
10. A completely empty
    space
11. A circular path
12. A force that pulls toward the
    center of the Earth
13. Force field
14. Negatively charged object

d. Inertia
e. Momentum
f. Orbit
g. Explosion
h. Rotation
i. Satellite
j. Speed
k. Gravity
l. Weight
m. Electron
n. Parallax
o. Mass
p. Magnetism

Part B   Choose your answer carefully.

1. A measurement of velocity includes   (a) force and time
   (b) speed and direction   (c) speed and time   (d) force and
   speed.
2. A change in velocity is called   (a) acceleration   (b) speed
   (c) force   (d) distance.
3. If you told a friend that an automobile was going 50 mi/hr, you
   were probably using   (a) the sun   (b) the moon   (c) the earth
   (d) Mars as a reference.
4. The   (a) pound   (b) ounce   (c) meter   (d) newton   is *not* a
   unit of force.
5. An object is in equilibrium when the result of the forces on it

(a) equal zero   (b) are unbalanced   (c) cause it to accelerate
(d) cause friction.

6. Friction is   (a) weight   (b) force   (c) mass   (d) velocity.

7. An object will accelerate when it is   (a) acted upon by an
   unbalanced force   (b) going at a constant velocity   (c) not
   moving   (d) in equilibrium.

8. Which of the following choices reminds you of Newton's second
   law?   (a) inertia   (b) $a = F/m$ (c) action-reaction
   (d) gravity.

✤ 9. An object whose mass is 10 kg is moving at a velocity of 5 m/sec.
   Its momentum is   (a) 0.5   (b) 2   (c) 15   (d) 50   kg-m/sec.

10. Marble A is at rest. An identical marble B is moving north and
    strikes marble A head-on in an elastic collision. After the
    collision   (a) marble A moves south   (b) marble B continues
    moving north   (c) marble B moves south   (d) marble B stops.

11. The path followed by a revolving object is called its   (a) orbit
    (b) velocity   (c) rotation   (d) axis.

12. The force that pulls objects toward the earth's center is
    (a) friction   (b) gravity   (c) momentum   (d) torque.

13. The amount of gravitational force between two objects depends
    on their   (a) mass and distance   (b) velocity and mass
    (c) electric charge and distance   (d) rotation and mass.

14. In a vacuum, heavy objects fall   (a) a little faster than
    (b) much faster than   (c) at the same rate as   (d) slower than
    light objects.

15. A rifle bullet is fired horizontally over level ground. At the same
    time, a bullet is dropped from the same height.   (a) The dropped
    bullet reaches the ground first   (b) The fired bullet reaches the
    ground first   (c) Both bullets reach the ground at the same time
    (d) There is no way to tell.

16. Weightlessness is experienced by someone in an earth satellite
    because   (a) there is no gravity acting on the satellite   (b) a
    person's mass becomes zero when leaving the earth's surface
    (c) a person is going so fast   (d) the astronaut and the satellite
    are in free fall.

17. One m/sec/sec is a unit of   (a) acceleration   (b) velocity
    (c) speed   (d) distance.

18. One newton is a unit of   (a) acceleration   (b) force   (c) speed
    (d) momentum.

19. All objects attract all other objects because of   (a) electrical
    (b) gravitational   (c) magnetic   (d) nuclear   force.

20. A force field is (a) a positive charge (b) an electron (c) a region of force (d) the North Pole.
21. An electron is (a) a magnet (b) a negative charge (c) a force (d) a pith ball.

Part C Think about and discuss these questions.

1. What is a frame of reference?

2. State Newton's three laws of motion in your own words.

3. Describe what condition causes equilibrium.

4. What is the difference between the mass and weight of an object?

RADIANT ENERGY: LIGHT AND WAVES

Light that "talks": telephoning by light beam may soon be possible! Researchers at Bell Laboratories have developed fibers, as thin as human hair, that can carry light signals. This photograph shows several stages that glass must undergo to be changed into a fiber. Now words that are spoken into a telephone are changed into electrical pulses which are passed along wires to the listener. Instead, the electrical pulses will be changed into light pulses that can pass through hair-thin semiconductor "wires". These light pulses will be changed back into voice signals in the telephone receiver. Think of the vast wire telephone systems now in use in this country. What a quantity of metal — a precious resource — will be "saved" by the use of fibers!

1 | INTENSITY AND ILLUMINATION OF LIGHT

YOUR OBJECTIVE: To find out what light is, and how it travels; to find out about the sources of light; to learn how the intensity and illumination of light are measured.

Light is a form of energy that makes it possible for you to see things. Isaac Newton thought that light was made up of *particles*. He likened a beam of light to a moving stream of particles. He thought that this stream of particles entering the eye made sight possible.

The Dutch scientist, Christian Huygens (1629–1695); thought that light was made up of *waves* instead of particles. Waves are like the ripples you see when you drop a pebble into a puddle of water.

Today, most scientists think that light is made up of both particles and waves. This idea was developed by the German scientist, Max Planck (1858–1947). It is called the *quantum theory*. **The quantum theory states that light is made up of particles (bundles) of electromagnetic waves.** Each particle is called a *photon* or *quantum*.

The direction in which light travels is represented by rays. In Figure 1-1, the rays of light are shown as straight lines and the arrowheads point in the direction that light travels from the source, the light bulb. Since light completely surrounds the bulb, the whole area could be

FIGURE 1-1

covered with light rays. But, in a drawing only a few rays can be used to show direction.

A material that allows light to pass through it is called transparent. A vacuum is perfectly transparent. Most materials are not perfectly transparent. A clear, thin pane of glass, clear plastic, or air are examples of materials that are almost transparent.

If a material is not perfectly transparent, then it either absorbs or reflects light rays, or blocks them completely. **Translucent materials allow some light rays to pass through and block others.** Frosted glass, plastic used for shower curtains, and cloth used for window curtains are examples of translucent materials. **Opaque materials block light rays entirely.** Materials made of metal, wood, plaster and most stones are opaque (ō-ˡpāk).

Energy The ability to do work.

Particle A piece of matter that is so small it cannot be seen by the naked eye.

Theory A general principle that relates various facts together.

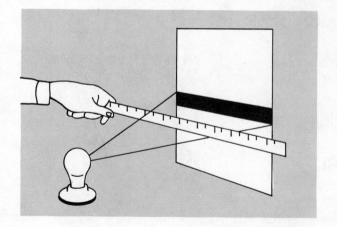

Absorbed light rays are rays that are stopped by a material. Opaque materials absorb all light rays. Translucent materials absorb some of them.

Reflected light rays are rays that strike an object and bounce off. You are familiar with a flat mirror that reflects light. Some materials, like water, reflect some light rays and allow others to pass through.

Refracted light rays are bent light rays. A magnifying glass has a lens that bends light in such a way that an object seen through it appears to be magnified, or made larger. The lenses of eyeglasses focus light rays into the eyes by refracting the light rays that pass through the lenses. Another example of a refracting material is water. An object under water seems to change shape. If you dip a ruler in a tub of water, the part of the ruler under the water looks as though it is bent where it passes through the surface of the water,

and the portion under water seems larger.

When light leaves a source of light, such as a light bulb or candle, it travels in a straight line. When an opaque object blocks the passage of light, **a shadow forms in the area blocked off from light by an opaque object.** If a very small light is used, a very dark, sharp shadow will be formed. (FIG. 1-2).

The shadow caused by any body in space that blocks the sun's light from any other body in space is called an *eclipse* (FIG. 1-3). **A solar eclipse, or eclipse of the sun, occurs when the moon's shadow is seen on the Earth's surface** (FIG. 1-4). The moon blocks the sun's rays from certain parts of the Earth, because the moon is passing between the Earth and the sun. **A total eclipse occurs when light from the sun is totally blocked from certain areas of the Earth by the moon's shadow.** At this time no one can see any part of the sun. However **during a partial eclipse,**

FIGURE 1-3 The sun fades in eclipse behind the black disk of the Earth as the Apollo 12 astronauts head for the second moon landing. *(NASA)*

FIGURE 1-4

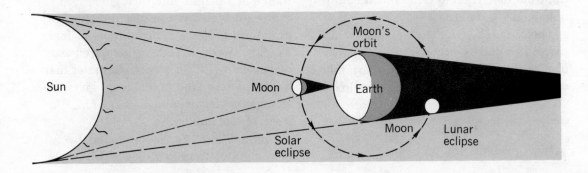

only a portion of the sun can be seen.

When the moon enters the shadow of the Earth, a lunar eclipse, or eclipse of the moon occurs (see FIG. 1-4). This means that the Earth is passing between the sun and the moon. The moon will become dark during such an eclipse.

Objects that are visible because they give off their own light are luminous. If an object becomes luminous when it is heated, it is incandescent. Candle flames and tungsten filaments (FIG. 1-5) in electric light bulbs are incandescent. The filament in a light bulb is heated by electric current passing through it. This, in turn, causes the bulb, or actually the filament, to become luminous and give off light.

Objects that become luminous when struck by invisible rays are called fluorescent. Since these objects do not give off heat when they become luminous, they are said to give off a "cold" light. For example, a television screen becomes fluorescent when it is showered with *electrons*. Fluorescent lamps give off light rays when struck by ultraviolet rays within the light tube. Fluorescent lamps are often used where bright light is required (FIG. 1-6).

Objects that do not give off their own light, but reflect it are illuminated. The moon can be seen because it reflects light from the sun. Therefore the moon is illuminated by the sun.

The amount of light given off by an object is called its intensity. The intensity of a powerful searchlight is certainly much greater than that from an ordinary

Electron An electrically charged particle in an atom.

flashlight. **The intensity of a light source is measured in units called candles:** 1 candle is the amount of light given off by 1 standard candle.

The *standard candle* used to be the amount of light given off by a candle made of a waxy substance that comes from whale oil. This candle was of a certain size and had to be burned at a certain speed. But now the standard candle is the *new candle.* The new candle is based on the light intensity that comes from the incandescence obtained by heating a ball of platinum to its melting point and directing the incandescent light through a hole 5 square millimeters wide. The light coming through this hole has an intensity of 1 candle.

A 40-watt tungsten filament light bulb gives off about as much light as 35 standard candles. Therefore, this bulb has an intensity of about 35 candles. A 40-watt fluorescent lamp gives off almost six times that much light, or about 200 candles.

The amount of light falling on an object is the illumination of the object. Again scientists needed a unit to measure the amount of illumination, or *illuminance* as it is sometimes called. **The unit of illuminance (illumination) is the footcandle:** 1 footcandle is the illuminance obtained at a distance of 1 ft from a light source of 1 candle intensity.

A light meter is used to measure illumination. It has a scale in units of footcandles. If you have a camera, you may be familiar with a type of light meter that has a different kind of scale — it gives the settings of the camera lens openings instead of footcandles.

Illumination can be increased by increasing the intensity of light. This can be

FIGURE 1-5 Notice the lighted filament in the large light bulb that the young woman is holding. The filament is a thin wire that glows when the light is turned on. *(General Electric)*

FIGURE 1-6 This young woman is holding two large fluorescent light tubes known as Econ-O-Watt lamps. These energy-saving lamps are designed to cut power needs up to 25 percent compared to conventional lamps. *(Westinghouse)*

done by replacing one light bulb with another one of higher intensity. If an object is moved closer to a light bulb, the illumination will surely increase. Illumination is also increased by adding more light bulbs to a lamp — some lamps hold several bulbs — or by using more lamps.

As light rays move farther away from most sources of light, they spread farther apart (see Fig. 1-1). And as light spreads over a greater area, the illumination lessens. If rays of light did not spread out, illumination would be the same at all distances from a light source. The light from a flashlight can be seen from a long way off. The light rays from the bulb in a flashlight spread out, but a curved mirror behind the bulb reflects these rays and brings them closer together so that they travel in parallel lines. If you were to take a flashlight into outer space, where there is a vacuum, you would be able to see its light from a distance of millions of miles. However on Earth the flashlight cannot be seen from such a great distance. This is because the air is not perfectly transparent. So the rays from the flashlight are reflected and absorbed as they pass through the air.

✳ Light spreads out from a light bulb in much the same way as paint spreads out from a spray can. The area covered by the light increases as the square of the distance from the light source. Because the light spreads out, the illumination decreases as the square of the distance — this relationship is called the *inverse square law* (Fig. 1-7). Therefore it has been determined that **the illuminance in footcandles equals the intensity of the**

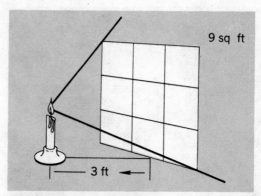

FIGURE 1-7 The light from a candle 3 feet from a screen produces a lighted area on the screen that is approximately 9 square feet.

light source in candles divided by the square of the distance in feet, or

$$Illuminance = \frac{Light\ source\ intensity}{(distance)^2}$$

This can be written in a shorter form,

$$I = \frac{L}{d^2}$$

✳ The intensity is constant for a given light source. A 100-watt light source will remain a 100-watt source no matter what the distances. Therefore the illuminance *I depends* on the intensity of the light source *L* and the distance (squared) d^2.

SAMPLE PROBLEM: What is the illuminance at a distance of 4 ft from a light source with an intensity of 160 candles?

SOLUTION: Using the above equation, *I* is the illuminance that you are asked to find. The light source intensity *L* is 160 candles and the distance *d* is 4 ft, so

$$I = \frac{L}{d^2} = \frac{160\ candles}{(4\ ft)^2} = \frac{160\ candles}{16\ ft} = 10\ footcandles$$

Measure the illumination from a light bulb.

Light meter
Yardstick
100-watt bulb
Light socket

Use a light meter to measure the illumination at various distances from a 100-watt light bulb. Take a meter reading at some distance with the 100-watt light on. Take a meter reading at the same place with the light off. This second reading will give you the illumination from light sources, other than the bulb, such as light coming through windows or from another room. The difference between these two readings will be the illumination given by the 100-watt bulb.

Part A

Point the light meter directly at the bulb from a distance of 1 ft. What is the illumination from the bulb? What is the intensity of the bulb?

Part B

Predict the illumination from the bulb at a distance of 2 ft. Check your prediction with the light meter. Was your prediction correct? Take meter readings at distances 3, 4, 5 and 6 ft. Chart the meter reading at each of these distances, including those at 1 and 2 ft, from the bulb: is the illumination at 2 ft half as much as it was at 1 ft? Is the illumination at 3 ft one third as much as it was at 1 ft? Can you explain your results?

**OBJECTIVE 1
ACCOMPLISHED?
FIND OUT.**

1. What is light? According to quantum theory, what is light made of?
2. Define the items reflected and absorbed as they apply to light rays.
3. Compare transparent, opaque and translucent materials. Give example of each.
4. What is the difference between an object that is luminous, and one that is illuminated?
5. What is the difference between a light source that is incandescent and one that is fluorescent?
6. What is a unit used for measuring the intensity of a light bulb? What is the unit used to measure illumination?
7. Why does illumination lessen as you get farther away from both light sources? Why does illumination change very little as you get farther away from a flashlight?
✳ 8. Suppose the intensity of a certain light bulb is 72 candles. Find the illumination at distances of 2 ft and 3 ft.

2 | HOW LIGHT IS REFLECTED

YOUR OBJECTIVE: To find out how images are formed by different arrangements of plane mirrors; to find out how images are formed by curved mirrors.

Objects that are not luminous can be seen only because they reflect light. When light rays strike a window pane, most of them pass through the glass. However, a small amount of light is reflected from, or bounces off, the surface of the glass. Reflected light makes it possible to see the reflection of objects in the glass. So the **light rays from an object are reflected, but not the object itself.**

A mirror reflects practically all of the light rays that strike it. Only a very small number of these rays are absorbed by the mirror. For this reason, the likeness of an object is much more clearly seen in a mirror than in a clear pane of glass.

When you see a likeness of an object in a plane mirror, it seems to be behind the mirror. This likeness is called the image. The object, itself, is called the object. Figure 2-1 shows the light from a light bulb being reflected in a mirror. Notice the object and the image.

What happens to the image as the object is moved farther from the mirror? The image also moves farther behind the mirror. **The image is always located the same distance behind a plane mirror as the object is in front of it.**

The image of an object seen in a plane

Plane mirror   A flat mirror.

mirror is reversed. If you are looking at yourself in a mirror and you raise your right hand, your image will appear to raise its left hand as it faces you. Just your right and left sides are reversed, not your head and feet. The reason your image seems to be reversed is only because it is *facing* you.

It would be very confusing for you if your image were not reversed in this way. You can prove this by placing two mirrors at right angles (FIG. 2-2). Then stand in front of the mirrors. You will see three images of yourself, and one of them is not reversed. Now if you raise your right hand, this image will appear to raise its right hand. But it would be difficult for you to use this arrangement of mirrors in your bathroom.

Two mirrors can be used in many ways to produce more than one image of the same object. The position of two mirrors will determine how many images are formed. **If two mirrors are placed at right**

FIGURE 2-1

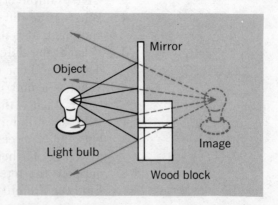

FIGURE 2-2 From left to right, the young woman's first image in the mirror at the far left is reversed — her left arm is the image's right arm. The next image is *not* reversed — the raised right forearm on the image corresponds to the woman's raised right forearm. The third image, at the right is again reversed — the woman's raised right forearm is seen in the mirror on the image's left side. *(David Strickler from Monkmeyer)*

angles, you will see three images (see Fɪɢ. 2-2). **If the mirrors face each other, and an object is placed between them, an *infinite number* of images will be reflected,** although you may be able to count only a few of them.

You are probably familiar with the common magnifying glass that makes objects look bigger when they are seen through the glass. The reason objects look bigger is because of the property of *magnification*. A mirror can also magnify if it is curved rather than flat (plane). **When an object is placed in front of a plane mirror, the image is always the same size as the object.**

> *Infinite number* A number of items is infinite if there is no end to them; that is, if you start to count them, you can never finish because there is always one more item.

FIGURE 2-3 Perhaps you have seen an image like this if you have visited a "fun house" at an amusement park. *(David Strickler from Monkmeyer)*

In curved mirrors, however, the image may be larger or smaller than the object. The magnification (or reduction) depends upon the amount of curve in the mirror and how far the object is from the mirror. The image may be upright or upside down. It also may be distorted (Fig. 2-3).

When the surface of a mirror is curved inward, the mirror is concave (Fig. 2-4a). When the surface is curved outward, the mirror is convex (Fig. 2-4b).

Parallel light rays that strike a concave mirror reflect toward a common point.

FIGURE 2-4

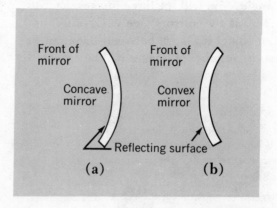

Front of mirror

Concave mirror

Front of mirror

Convex mirror

Reflecting surface

(a)   (b)

Compare the position of an object and its image.

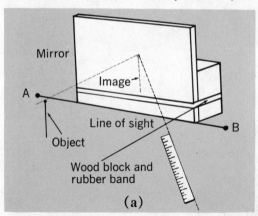

(a)

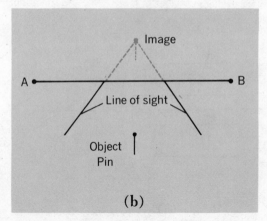

(b)

Plane mirror
Wooden block
Rubber bands
Straight pins
Paper
Cardboard
Metric ruler

Place a sheet of paper on top of a piece of cardboard. Draw a line AB near the center of the paper. Attach one edge of the plane mirror to a wooden block with a rubber band so that its reflecting surface is facing you. This will keep the mirror in an upright position. Place the bottom edge of the plane mirror (attached to the block) on the paper so that its reflecting surface is perpendicular to line AB as in Figure (a). Stick a pin in the cardboard about 5 centimeters from the front of the mirror. This pin will serve as an object whose image will be located in the mirror.

Lay a ruler on the paper on one side of the object pin. Look along the edge of the ruler at the image of the pin in the mirror. Draw a *line of sight* along the edge of the ruler while it is in this position. (A "line of sight" is an imaginary straight line from your eye to an object.) In the same way, draw another line of sight on the other side of the object pin. Remove the mirror. Continue the two lines of sight by drawing dotted lines until they meet in back of line AB as in Figure (b). Line AB now replaces the mirror. Where is the image located? Measure the distances of the object and the image from line AB. How do these distances compare?

This point, at which the rays cross, is a focus or focal point. The location of a focal point depends on the direction at which the parallel rays strike a mirror (FIG. 2-5).

The principal axis is an imaginary line passing through the center of a curved mirror at right angles to the mirror. When light rays strike a concave mirror

Axis An imaginary line passing through the center of an object.

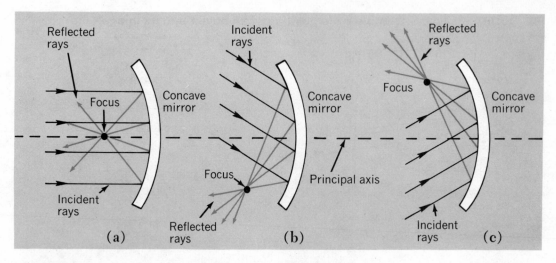

FIGURE 2-5 (a) Notice the principal focal point. In (b) and (c) the parallel rays strike the concave mirrors from different angles. This causes the focal points to be in different locations.

parallel to the principal axis, they reflect and cross at a focal point called the **principal focal point** (see FIG. 2-5a).

When an object is placed in front of a concave mirror, two different types of images can be formed. **An object that lies beyond the principal focal point has a real image. A real image is formed by light rays that focus at a point.** If a screen is placed at that point, the image can be seen on the screen, but it will be upside down. This happens because a ray from the top of an object will pass through the focal point to the lower portion of the reflecting surface of the mirror, and a ray from the bottom of the object will pass through the focal point to the upper portion of the mirror's surface (FIG. 2-6).

An object that lies between the principal focal point and the mirror has a vir-

FIGURE 2-6 The real image is upside down on the screen behind the object.

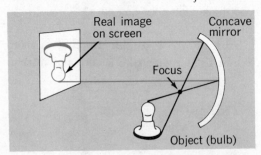

FIGURE 2-7 The virtual image is right side up.

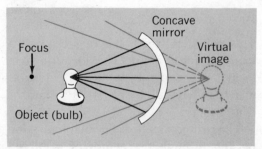

Part A

2 mirrors
2 rubber bands
2 wooden blocks
Small object

Stand two mirrors against wooden blocks, place them at right angles (see FIG. 2-2). Place an object in front of the mirrors. How many images do you find? Now change the angle between the mirrors a small amount, but keep them touching on one edge. How many images can you count? Now change the angle between the mirrors again but still keep them touching. At what angle of the mirrors do you get the largest number of images? What happens to the number of images as the angle between the mirrors is widened?

Part B

Place an object between two mirrors that face each other. How many images can you count? Explain.

tual image. The virtual image is *behind* the mirror (FIG. 2-7). It is located in the place from which the reflected rays appear to come. It is the same type of image that forms in a plane mirror except that it will be magnified if the mirror is concave. **A virtual image cannot be focused onto a screen like a real image can because the rays of light do not come to a focus.** However, a virtual image can be seen.

When an object is at the principal focal point, there will be no image. The rays from the object are reflected parallel to each other (FIG. 2-8). Searchlights, automobile headlights, and flashlights make use of concave mirrors in this way to send out a beam of parallel rays.

When parallel rays of light strike a convex mirror, the reflected rays spread out (FIG. 2-9). Because the rays that strike

FIGURE 2-8   The object (the light bulb) is *at* the focal point.

FIGURE 2-9

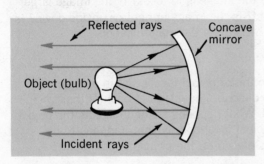

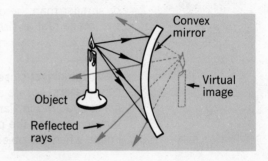

a convex mirror do not focus at a point, **real images cannot be formed.** Therefore only virtual images can be seen in a convex mirror.

Convex mirrors reduce the size of the image. This makes them useful in places where a person wishes to see a large area. They are often used in stores where clerks need to see a large portion of a store at a glance. For this reason convex mirrors are used as rear view mirrors on school buses and other large vehicles.

BE CURIOUS 2-3: **Find the magnification and location of images in a curved mirror.**

Part A

Paper screen
Concave mirror
Convex mirror
Wooden block
Rubber band
Masking tape
Metric ruler
Candle (or small lightbulb with wires and battery)

Use a concave mirror to focus a distant object (such as a distant building) on a paper screen. Measure the distance from the image on the screen to the mirror. This distance is the *focal length* of the mirror. Check your results by focusing other distant objects on the screen.

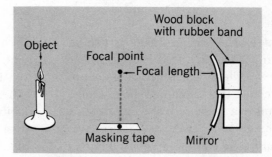

Part B

Stand the mirror on a table as shown in the figure. At a distance of one focal length in front of the mirror, put a dot on a piece of masking tape stuck to the table. The dot marks the point on the table which is directly below the principal focal point. Place an object (a candle or small light bulb) about two meters in front of the mirror. Focus the image of this object onto a screen. Where is the screen located when the image is in focus? Is the image larger or smaller than the object? Is the image upright or upside-down? Is the image real or virtual? Estimate the magnification.

Part C

Move the object closer to the principal focal point in steps of 25 centimeters. Locate and describe the image at each of these steps just as you did in Part B. What happens to the image as the object moves closer to the principal focal point? What happens to the image when the object is placed on the principal focal point?

Part D

Place the object between the principal focal point, and the mirror. Can the image be focused on a screen? Is the image larger or smaller than the object? Is it upright or upside down? Is this a real or virtual image? Slowly move the object closer to the mirror. What happens to the image?

Part E

Replace the concave mirror with a convex one. Place the object about two meters in front of this mirror. Describe the image formed in the mirror. Move the object closer to the mirror in steps of 25 centimeters. Look into the mirror and notice the image at each of these steps. What happens to the image as the object moves closer to the mirror? Describe the image when the object reaches the mirror.

OBJECTIVE 2 ACCOMPLISHED? FIND OUT.

1. What is reflected in a mirror? What is an image?
2. When an object is moved farther from a plane mirror, what happens to the image? What is the position of the image compared with the position of the object?
3. How many images of an object will be produced in two mirrors set at right angles to each other? How many of these images will be reversed?
4. How many images will be produced if an object is placed between two parallel mirrors?
5. What happens to the direction of parallel rays of light after reflection from a convave mirror? After reflection from a convex mirror?
❋ 6. Why is a light bulb placed at the principal focus of the concave reflector in a flashlight?
❋ 7. Where should an object be to produce a real image in a concave mirror? Where should it be to produce a virtual image?

3 | REFRACTION OF LIGHT

YOUR OBJECTIVE: To find out what refraction of light means and what causes refraction; to find out how white light can be separated into colors by prisms and filters.

FIGURE 3-1 This photo shows a straw in a glass of water. Notice that the straw looks as through it were broken where it enters the water. This is an example of the refraction of light. *(Mimi Forsyth from Monkmeyer)*

Have you ever looked at a straight stick when part of it was submerged in water? It appears as if the stick were bent at the point where it enters the water. You know that the stick is not really bent because when you take it out of the water it is still straight.

Looking at the stick through the water gives it the bent appearance. This is because the water changes the path of the light rays that travel to your eye from the part of the stick that is underwater. The light bends or changes its path as it leaves the water and enters the air. **Light can be bent when there is a change in its speed as it passes from one material to another. This bending of light is called refraction.** Light travels at different speeds through different materials. It is greatest when it passes through a *vacuum*. The speed is slightly less when light travels through air, and still less when it passes through substances, such as glass and water (FIG. 3-1). Therefore, refraction occurs when light travels from air into glass, from a vacuum into air, or from water into glass, and so on. It can even be seen when light passes from warm air into cold air.

The position of objects seems to change because of refraction. A fish in a pond can be seen because light travels to the eyes from the fish (FIG. 3-2). The light rays coming from the fish refract as they leave the water. This makes the fish look like it is in a different place.

> *Vacuum* A completely empty space.

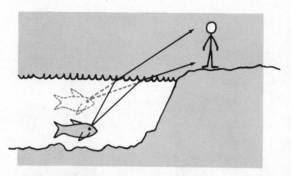

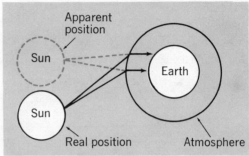

FIGURE 3-2 To the stick figure the fish appears to be in the position shown by dotted lines. The actual position of the fish in the pond is shown by solid lines.

FIGURE 3-3

Because of refraction, the sun while rising or setting seems to be higher in the sky than it really is (Fɪɢ. 3-3). The sun's rays bend as they enter the air layer that covers the Earth. They are slowed down as they pass from the vacuum in outer space into the air. People on Earth then see the rising or setting sun as higher in the sky than it really is.

Many *optical instruments* make use of refraction. Some of these are telescopes, binoculars and periscopes. These instruments have special lenses that refract light rays.

In the seventeenth century, Isaac Newton found that white light rays from the sun could be separated into various colors after passing through a triangular piece of glass called a prism (Fɪɢ. 3-4). He saw a band of colors ranging from red to violet — red, orange, yellow, green, blue and violet. From this Newton was able to say that white light contains all these colors. In fact, the common rainbow that results from sunlight passing through tiny drops of water happens because the drops of water act like prisms.

A glass prism separates the colors of light because each color is refracted a different amount by the glass. The prism refracts or bends violet light the greatest amount and red light the least amount. The colors separated by a prism are called the *visible spectrum* because they

Optical instrument A device used for seeing things, such as a microscope or telescope, or even a simple magnifying glass.

Binoculars An optical instrument that magnifies distant objects, which can be seen with both eyes.

Periscope An optical instrument containing mirrors or reflecting prisms which are placed at each end of a vertical tube. A person looking into a periscope can see objects behind a wall or overhead.

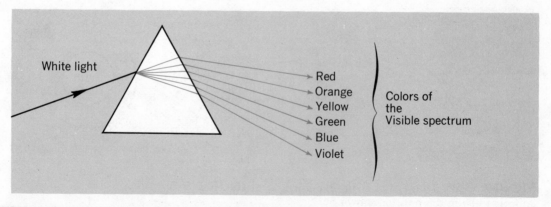

FIGURE 3-4 White light enters triangular glass prism where it is refracted into a band of the six colors of the visible spectrum.

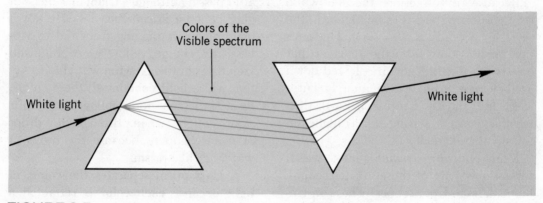

FIGURE 3-5

can be seen. The colors of the visible spectrum can be passed through another triangular glass prism that bends them back into white light again (FIG. 3-5).

A piece of red glass will allow only red light rays to pass through it. This glass stops all colors other than red from passing through it (FIG. 3-6). It is opaque to all other colors. Similarly, blue glass allows only blue light to pass through it, and so on, with every possible color.

FIGURE 3-6

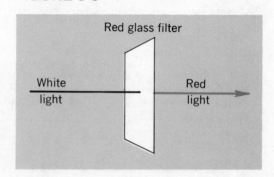

A device that passes only one element and blocks all others is called a filter. Therefore, the red glass is a filter because it passes only red light rays and blocks rays of other colors. You have probably heard of water filters that remove impuri- ties from water. The water filter does this by preventing impurities from passing through it.

When a blue filter is placed behind a red filter that receives white light, no light will pass through the blue filter (FIG.

BE CURIOUS 3-1:   **Observe the refraction of light.**

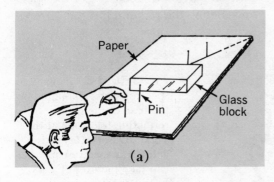

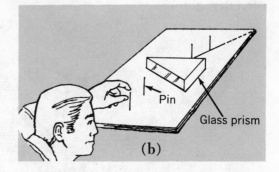

Part A

Transparent rectangular block of glass

Paper

Cardboard

4 pins

Protractor

Triangular glass prism

Place a transparent rectangular block of glass near the center of a piece of paper mounted on cardboard as shown in Figure (a). Use a pencil to trace the outline of the block on the paper. Draw a line from the center of the outline of the glass block to the right-hand corner of the paper. Place two pins on this line about an inch apart. Look straight through the glass block from one side of the paper at the two pins. Now move until you can see the two pins along the line to the right-hand corner of the paper, one in back of the other. Place two pins between your eye and the glass block along this same line of sight. All four pins should now look as if they were lined up by your eye as it looks through the glass block. Remove the glass block. Can you show how the line of sight bent as it entered and left the glass block?

✿ **Part B**

Repeat Part A using a triangular glass *prism* in place of the glass block as shown in Figure (b). (A prism is a transparent object that bends light rays. It has two surfaces that are not parallel, and is usually made of glass or clear plastic.) By moving the pins, can you find a point at which the pins cannot be seen through the prism? If so, can they be seen through the edge along the bottom of the prism? Explain.

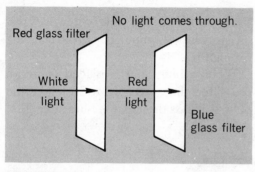

FIGURE 3-7

3-7). This is because the only light that strikes the blue filter is the red light that passes through the red filter. And the red light will not pass through the blue filter. It will be completely absorbed.

❋ The colors that you see in the spectrum band are called the "visible" spectrum. However, the spectrum continues into an *invisible* portion at each end of the spectrum. Below the red end lies the *infrared*

FIGURE 3-8 This spectrograph is mounted on the base of the giant refracting telescope at the Yerkes Observatory It is actually a spectroscope that can take pictures of stars. In this spectrograph, light goes through three prisms which spread the light into colors so that the star's spectrum can be photographed. A star's spectrum aids astronomers in determining the temperature and composition of a star. It also enables them to determine the velocity at which the star is traveling toward or away from them. (*Yerkes Observatory*)

(in-fra-'red) portion of the spectrum. Above the violet end lies the *ultraviolet* portion. Both the infrared and ultraviolet portions of the spectrum are invisible.

Light travels in waves. Each wave has a different length. Because of this you see a particular color — red, blue, yellow, etc. The color at the lower end of the spectrum is red. This color has very long waves as compared with the short waves that make up the violet at the upper end of the spectrum. When a wave of a certain length is refracted by a prism, a particular color is seen. This property — that colors can be separated by a prism — enables astronomers to tell how hot a certain star is or what gases make up a star.

They find out these things by using a telescope that has a spectroscope (FIG. 3-8). A spectroscope is an optical device that has a prism which refracts light. If all colors are present in the light source, an observer viewing the light through the spectroscope will see a complete rainbow pattern. If a color is missing, the observer sees a dark line at the position of the missing color. If some color is present in a very large quantity, a bright line is seen in that place. The telescope magnifies a star and the spectroscope shows the colors of the light from the star. These colors give an astronomer information about the temperature of the star and the elements that make up the star.

OBJECTIVE 3 ACCOMPLISHED? FIND OUT.

1. What is refraction of light?
2. Under what condition can light be refracted?
3. Name several optical instruments that make use of refraction.
4. Why are the colors of the spectrum separated as they pass through a prism?
5. Complete the following sentence: A green filter will allow _____ to pass through it and it absorbs _____ .
6. What is the visible spectrum? Describe the invisible part of the spectrum.
7. What is a spectroscope, and why do astronomers use it?
✳ 8. What portion of the invisible spectrum lies at each end of the visible spectrum?

4 | REFRACTION WITH LENSES

YOUR OBJECTIVE: To find out how lenses refract light; to find out how lenses make images; to learn how simple optical instruments are constructed.

In Section 2 you found that curved mirrors could focus light rays at a point called the *focal point*. You also saw that curved mirrors could spread light rays. A lens is able to do these same things with light. The lens, however, focuses or spreads light rays by *refraction* instead of reflection.

A lens is a transparent material with at least one curved surface. Lenses are usually made of clear glass or plastic. **A convex lens is thicker at the center than at the edge** (FIG. 4-1a). **A concave lens is thicker at the edge than at the center** (FIG. 4-1b). Recall from Section 2 the terms *concave* and *convex* were used for mirrors.

A convex lens causes light rays to focus (FIG. 4-2a). **However a concave lens causes light rays to spread out** (FIG. 4-2b).

Parallel light rays that pass through a convex lens focus at a focal point (see FIG. 4-2a). **Parallel rays that pass through a concave lens are spread out by the lens** (FIG. 4-2b). The position of the focal point depends on the direction at which the light strikes the lens (FIG. 4-3).

Just as with a mirror, a lens has a *principal axis*. This is an imaginary line that is

FIGURE 4-1

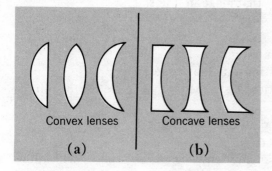

Convex lenses   Concave lenses

(a)   (b)

FIGURE 4-2

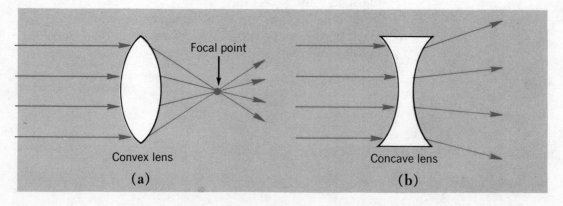

Focal point

Convex lens   Concave lens

(a)   (b)

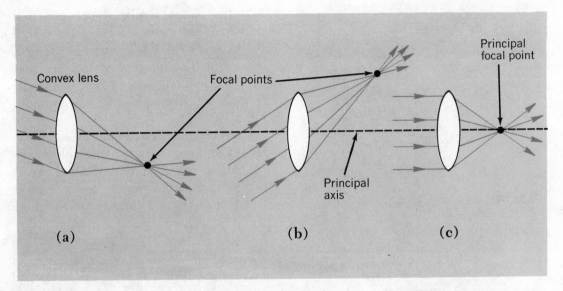

Convex lens

Focal points

Principal
focal point

Principal
axis

(a)   (b)   (c)

FIGURE 4-3

perpendicular to a lens and passes through its center. When the rays come in parallel to the principal axis, they focus at a point called the *principal focal point* (FIG. 4-3c). There is a principal focal point on either side of a convex lens.

If an object is placed beyond the principal focal point of a convex lens, there will be a real image (FIG. 4-4). This means that the image can be focused onto a screen, and it will be upside down.

If an object is at the principal focal point of a convex lens, no image is formed (FIG. 4-5). Compare this with the results of an object placed at the focal point of a concave mirror discussed in Section 2.

FIGURE 4-4

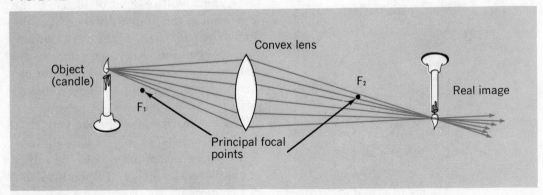

Object
(candle)

Convex lens

F_2

Real image

F_1

Principal focal
points

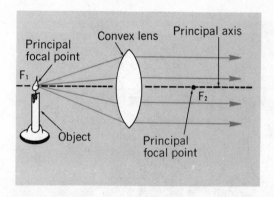

FIGURE 4-5

If the object is placed between the principal focal point of a convex lens and the lens, a virtual image will be formed (FIG. 4-6). The image is virtual because the light does not focus in this region, but spreads out. This spreading causes magnification.

Cameras make use of a movable convex lens. The object being photographed lies beyond the principal focal point, so a real image is formed. A camera uses a convex lens to focus a real image onto the film

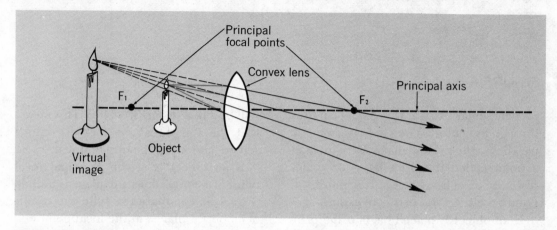

FIGURE 4-6

FIGURE 4-7

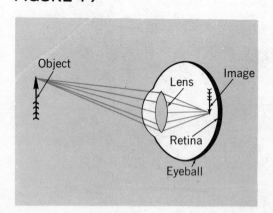

inside the camera. The picture taken by the camera will be "out of focus" if the lens is not at the proper distance from the film. When an object is far from a camera, the lens is moved closer to the film to get a sharp image on the film. When an object is near the camera, the lens is moved farther from the film.

The eye works in a similar way to a camera (FIG. 4-7). The lens in the eye must focus a sharp image onto the retina inside the eye, which corresponds to film in a camera.

BE CURIOUS 4-1:   **Find out how a convex lens magnifies and where the image will be.**

Convex lens
Clay
Masking tape
Metric ruler
Paper screen
Candle (or small light bulb with wires and battery)

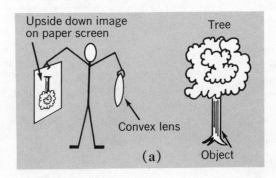

(a)

Part A

Use a convex lens to focus a distant object onto a paper screen as shown in Figure (a). Measure the distance from the image on the screen to the lens. This distance is called the *focal length* of the lens. Use the same method to find the focal length on the other side of the lens. How do the two focal lengths compare?

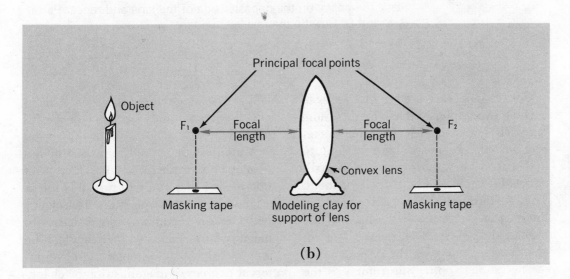

(b)

Part B

Stand the lens on a table top as shown in Figure (b). Stick two pieces of masking tape on a table at a distance of one focal length on both sides of the lens. Put a dot on each piece of tape to mark the point directly below the two principal focal points. Place an object (a candle or small light bulb) about two meters from the lens on either side. Focus its image onto a screen. Where must the screen be placed to get the image in sharp focus? Is the image upright or upside-down? Is it larger or smaller than the object? Is the image real or virtual?

Part C

Move the object closer to the nearest principal focal point in steps of 25 centimeters. Notice the image at each of these steps. What happens to the image as the object moves closer to the principal focal point? What happens to the image when the object is placed at the principal focal point?

Part D

Place the object between the principal focal point and the lens, starting at a point nearer to the focal point. Can the image be focused on a screen? Is the image larger or smaller than the object? Is it real or virtual? Slowly move the object closer to the lens. What happens to the image?

Part E

Place the object on the opposite side of the lens and repeat Parts B, C and D. How do the results compare?

Convex lenses are used in eyeglasses to correct hyperopic, or farsighted, vision. Similarly, **concave lenses are used to correct myopic, or nearsighted, vision.** In both cases the structure of the eyeballs causes them to be slightly out of focus. This, in turn, causes blurred images on the part of the retina at the back of the eyeball (Fig. 4-8). A corrective lens focuses a sharp image on the retina.

The simplest optical instrument is the ordinary magnifying glass. This instrument consists of a convex lens attached to a holder. When an object is held near the lens — between a focal point and the lens — the image appears larger. A magnified virtual and upright image is formed.

A microscope is used to make tiny unseen objects visible (Fig. 4-9). The type of microscope that you will probably use in a laboratory when you study life science will contain a combination of lenses to magnify tiny objects so that they can be seen clearly. A microscope can have different degrees of magnification by changing the lenses.

A telescope is used for seeing objects at great distances. It gives astronomers much useful information about objects in

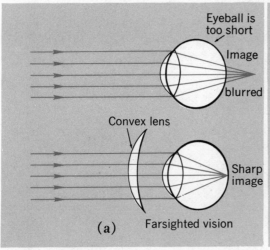

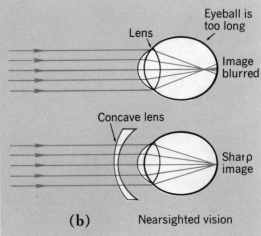

FIGURE 4-8 (a) Hyperopic, or farsighted, eyeball: Light rays focus behind retina, so convex lens (thick at center) is used to bend light rays so that they focus onto the retina at the back of the eyeball. (b) Myopic, or nearsighted eyeball: Light rays focus before retina, so concave lens (thin at center) is used to spread light rays slightly so that they focus onto the retina.

FIGURE 4-9 This laboratory microscope has two lens eyepieces that can be used in combination with one of four additional lenses, which are mounted on a disk that can be turned to select any one of these lenses. (*Bausch & Lomb*)

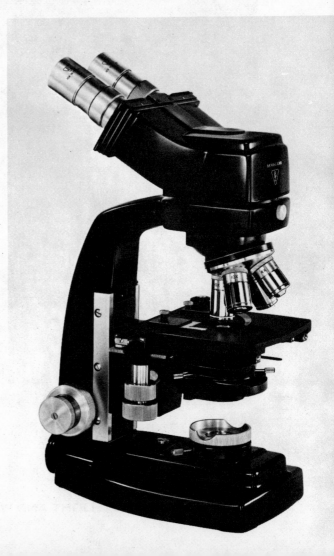

outer space. Telescopes also make use of various combinations of convex and concave lenses, as well as mirrors. *Refracting telescopes* have two main lenses. Those owned by amateur astronomers usually have lenses that are only a few inches in diameter. The world's largest refracting telescope is at the Yerkes Observatory in

FIGURE 4-10 The telescope at Yerkes Observatory. A spectrograph can be mounted on the base (refer to Fɪɢ. 3-8). *(Yerkes Observatory)*

Williams Bay, Wisconsin (FIG. 4-10). It has a main lens that is 40 inches in diameter. *Reflecting telescopes* use mirrors as well as lenses. They are usually larger than the refracting type. Amateur astronomers often build their own reflecting telescopes with mirrors that are 6 inches or larger in diameter. The Hale telescope at the Mount Palomar Observatory in California is the largest reflecting telescope. It has a main lens that is 200 inches in diameter.

BE CURIOUS 4-2: **Find out how some common optical instruments work.**

Magnifying glass
Binoculars
Box camera (or other optical instrument)

Look at a magnifying glass, binoculars and a simple box camera. Then make a simple sketch of the inside of each of these items showing how the lenses are set up for getting real or virtual images. Explain how your drawings work.

OBJECTIVE 4 ACCOMPLISHED? FIND OUT.

1. What is a lens?
2. What does a convex lens do to parallel light rays? What does a concave lens do to parallel light rays?
3. Where should an object be placed to produce a real image from a convex lens? A virtual image? Parallel light rays? (You can use a magnifying glass to answer these questions.)
4. What is the purpose of the lens in a camera? In your eye?
5. What type of lens is used to correct farsightedness? Nearsightedness?
6. How does a refracting telescope differ from a reflecting telescope?
* 7. Why is the film in a movie projector placed beyond the principal focal point of the projection lens?

5 | WAVES

YOUR OBJECTIVE: To find out what a wave is and how different types of waves are produced; to learn some of the terms used to describe waves; to find out how waves can be reflected, refracted, diffracted, and polarized; to understand the electromagnetic spectrum.

Christian Huygens pictured light as a wave of energy. He found that he could

FIGURE 5-1 This device is a ripple tank. Ripple tanks are used to study wave motion. This illustrates that energy from the splash point is passed along in all directions. (*Bond Manufacuring Company*)

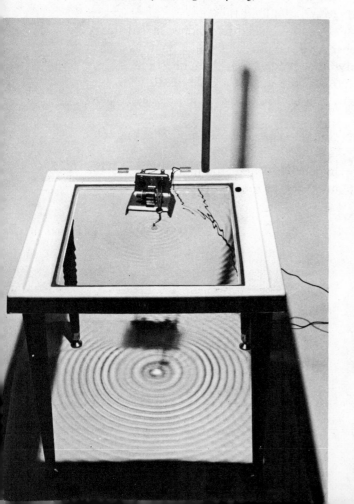

explain most of the properties of light if it could be thought of as waves.

A good example of a wave is the water wave, because it is probably the most familiar to most people. A water wave may start from some point where the water particles have been disturbed, or set into a vibrating motion by a force given to these particles. If you drop a rock on the calm surface of a pond you will create waves. You see these waves as a series of circles that move away from the point where the rock entered the water. That point is at the center of all these circular waves.

You can picture what is happening in this way. The rock pushes water particles downwards as it enters the water. The water particles now tend to return to their original position once the rock has gone down to the bottom.

This up-and-down motion is the result of the energy given to the water particles at the splash point. Some of this energy is passed on to water particles that surround the splash point, and they, in turn, begin to vibrate up and down. As this continues, the energy is passed along in all directions away from the splash point. You can see this as a series of circles that spread away in all directions (FIG. 5-1).

You can also see that the waves made by tossing a pebble into a pond travel — they move outward from the splash point. This type of wave is called a *transverse wave* because it moves "across." (The syllable *trans* means "across" in Latin.) **A transverse wave travels at right angles to**

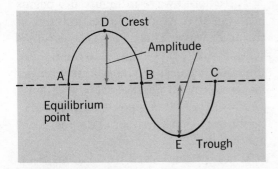

FIGURE 5-2   A model of a single wave.

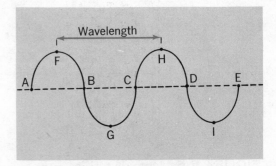

FIGURE 5-3   A model of two waves — two crests and two troughs. What two other pairs of points in this figure can be used to measure the wavelength?

the direction of the wave's disturbance. Thus the waves made by dropping the pebble *down* into the pond move *across* the pond. **A longitudinal wave travels in the same direction as the source of the wave's disturbance.** For example, if a coil spring is stretched and released, it will move in a back-and-forth motion to and from the point at which the spring was released.

Usually waves are not smooth, but Figure 5-2 will give you a general idea of a side view of a wave. In the figure point D is the *crest,* or top, of the wave and point E is the *trough,* or lowest point. **A full wave has a crest and a trough.** Points A, B, and C are *equilibrium points.* **The amplitude, or wave height, is the perpendicular distance from a line passing through the equilibrium points to either a crest or trough.**

The wavelength is the distance measured from one point on a wave to the corresponding point on the following wave. In Figure 5-3 the wavelength can be measured from point F to point H or

from point A to point C or from point G to point I.

The frequency is the number of full waves — that is crest and trough — that occur in a certain length of time.

The frequency is measured in *cycles per second* (cps). One cycle per second is also 1 *hertz,* named for the German physicist, Heinrich Hertz (1857–1894). A full wave is called a *cycle* because it is like a circle. Points A, B, C, D and E on the wave in Figure 5-4a are also shown as corresponding points on a circle in Figure

FIGURE 5-4   (a) Points on the wave model correspond to (b) points a circle.

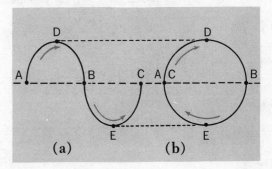

(a)   (b)

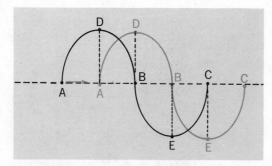

FIGURE 5-5 Phase-shift model — the wave shown in color — is shifted 90 degrees to the right.

5-4b. (The word *cycle* comes from a Greek word meaning circle.) **The period is the time it takes to complete one full cycle.** Some waves do not complete cycles so they are not *periodic*.

✻ Just as in a circle a wave can be measured in degrees. In Figure 5-4a and b point A is 0 degree, point D is 90 degrees, point B is 180 degrees, point E is 270 degrees and point C is 360 degrees — a full circle has 360 degrees. Notice that points A, B and C are at the same level. These points are *equivalent*.

✻ The colored wave in Figure 5-5 begins at its point A which is under point D of the other wave. This colored wave represents a "shift" of 90 degrees to the right of the other wave. It is *out of phase* with the first wave by 90 degrees. This is called a *phase shift* of 90 degrees. One wave may be shifted by any number of degrees from another wave.

When water waves travel from deep water into shallow water they slow down when they enter the shallow water region. When the speed at which the waves travel is changed (slowed) they are bent. **Waves are refracted when their speed changes.**

If straight waves pass through a narrow opening, the waves are *bent and spread out* as they pass through the opening. This action of waves is called *diffraction*. (FIG. 5-6). **Diffraction — bending and spreading out of waves — occurs when waves pass through openings smaller than one wavelength.** If water waves pass through

Equivalent Having equal amount or value.

FIGURE 5-6 Water waves pass through openings of three sizes. (c) Notice that the smallest opening at the right produces the most diffraction.

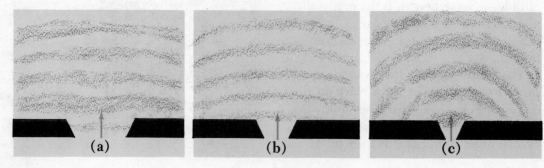

(a) (b) (c)

BE CURIOUS 5-1:   Use a simple method to study wave characteristics.

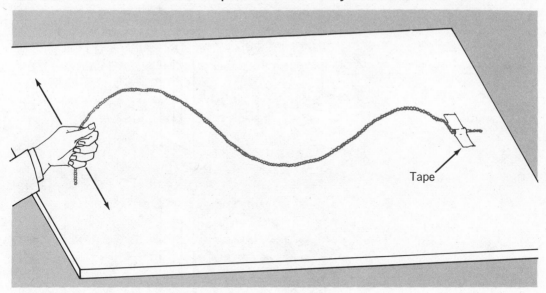

Tape

Beaded chain
Masking tape
Watch or clock with
second hand

Lay about two feet of beaded chain on a table top. Tape one end down. Practice sending a series of waves along the chain by moving your hand quickly to the left and right as shown in the figure. Once you can do this smoothly, make waves at a frequency of about one wave per second (1 hertz). Note the wavelength. Now increase the frequency by moving your hand at a faster rate. What happens to the wavelength as the frequency increases? How can you change the amplitude of the wave? In what direction do the beads move? In what direction does the wave move?

an opening *larger* than one wavelength, no diffraction will occur.

Sometimes waves can form interesting and useful patterns because of diffraction. Such *diffraction patterns* can be used to

calculate the wavelength of light. X rays directed at *crystals* form diffraction patterns. By studying such patterns, scientists can learn much about the atomic structure of crystals.

Crystal A regularly shaped, geometric arrangement of atoms in minerals.

Vertical Perpendicular to the plane of the horizon.

FIGURE 5-7

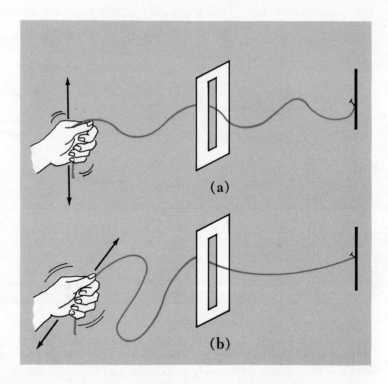

(a)

(b)

Suppose you make a *vertical* wave pass along a string through a vertical slot (FIG. 5-7). The wave will travel through the slot. What if you were to try sending a *horizontal* wave along the same string through the vertical slot? The vertical slot will prevent a horizontal wave from passing through it. Only a vertical wave will pass through the vertical slot. This slot acts as a filter that passes only vertical waves. Thus, the slot *polarizes* the waves vertically.

Similarly a horizontal slot will polarize waves in a horizontal direction. Waves can be polarized in any direction.

Certain materials polarize light waves that pass through them. These materials are called *polarizers*. They act much like the slots in Figure 5-7. When placed in a certain way, **a polarizer will allow only waves that are in the same plane as the polarizer to pass through it.** When the polarizer in Figure 5-7 is rotated 90 degrees (one quarter turn), only waves polarized in a horizontal direction will pass through.

Experiments have shown that the light reflected from a smooth surface tends to be polarized. This reflected light can cause a dangerous glare for the driver of an automobile on a bright sunny day. To cut down on this glare the driver may wear sunglasses that are polarizers. The polarizers are set in such a way that they greatly reduce the amount of reflected light reaching the driver's eyes.

In the latter part of the nineteenth century, James Clerk Maxwell (1831–1879), a Scottish scientist and mathematician, showed that **light is a combination**

Find out if you can produce diffraction patterns from light.

Showcase lamp
Bulb socket
Microscope slide
2 razor blades
Black liquid (graphite
in alcohol)

Coat one side of a glass microscope slide with the black liquid supplied. After the coating has dried thoroughly, make a single narrow scratch on it with a razor blade. Look at the light source through this narrow slit. Make an accurate drawing of the pattern in the light you see.

Now make two narrow scratches close together on the black coating. Look at the light source through the narrow slits. Compare this diffraction pattern with the one made by the single slit. Do you notice any difference?

One of the newest and important tools used in science and industry is the *laser*. The word laser stands for "**l**ight **a**mplification by **s**timulated **e**mission of **r**adiation." The laser sends out a beam of light made by some source like a small ruby rod.

A laser sends out a very narrow beam of light. Because the beam is so narrow, all of its energy is put into a very small area. The beam can be made more narrow by focusing it with lenses and mirrors. This makes the laser a powerful tool. The laser beam can be made so powerful that it can melt or vaporize many materials. A high-powered laser beam can pierce a razor blade, as shown in the photo.

A laser beam remains narrow even after traveling great distances. This makes it useful for measuring great distances accurately.

The Apollo astronauts left a mirror on the moon for just this purpose. A laser beam from the Earth was aimed at the mirror. It was reflected and returned to Earth. The information from this experiment was used to measure the distance between the Earth and the moon far more accurately than had ever been done before.

Scientists are constantly finding new uses for the laser. These new uses are in many areas including communications, safety and medicine.

ELECTROMAGNETIC SPECTRUM

Radio Waves	Infrared Light	Visible Light	Ultraviolet Light
Radio waves are at the lowest end of the electromagnetic spectrum, just below the infrared region. Therefore, radio waves have a lower frequency (longer wavelength) than infrared waves. Radio waves are used to carry radio, television and radar signals. Radio telescopes receive radio signals from stars and convert them into visible information.	The infrared (in-fra-'red) region is just below the visible spectrum. Thus infrared waves have a lower frequency (longer wavelength) than red light. Because infrared waves give off heat, they are used in heating elements in ovens. Infrared waves are also used for medical purposes.	Light makes up only a small portion of the electromagnetic spectrum. This is the *visible* portion of the electromagnetic spectrum. Red light has the lowest frequency (longest wavelength). Thus it is at the lowest end of the visible spectrum. Violet has the highest frequency (shortest wavelength), so it is at the uppermost end of this spectrum.	Ultraviolet waves are just above the visible spectrum. Thus these waves have a higher frequency (shorter wavelength) than violet light. Ultraviolet radiation is sometimes called "black light." When it strikes certain fluorescent substances, it makes them glow. Geologists use ultraviolet light to find certain fluorescent minerals.

of electric and magnetic waves. **This type of wave is now called an electromagnetic wave.**

Besides light, there are other types of electromagnetic waves. These other types are all *invisible*. The chart on these two pages shows the different types of elec-

tromagnetic waves. Those on the left side of the chart have the lowest frequencies (or longer wavelengths). The frequencies increase (or wavelengths shorten) as you move to the right along the chart. The entire range of electromagnetic waves is called the *electromagnetic spectrum*.

BE CURIOUS 5-3:   **Find out if various light sources cause polarized waves.**

Part A

Light bulb
Lamp socket
2 polarizing filters

Look directly at a light bulb through a single polarizing filter. Slowly rotate the filter. What is the result? Look through the filter at light reflecting from smooth surfaces around the room. Slowly turn the filter while looking at the reflected light. What do you see? Turn the filter while looking at different parts of the sky, including the clouds (do *not* look directly at the sun). What are your conclusions?

Part B

Look at a light bulb through two polarizing filters in line with the bulb. Slowly turn one of the filters while holding the other one fixed. What happens when the filter has made one full turn? Explain.

ELECTROMAGNETIC SPECTRUM (continued)

X Rays	Gamma Rays	Cosmic Rays
X rays are above the ultraviolet region of the electromagnetic spectrum. The frequency is higher and the wavelength is shorter than ultraviolet waves. It was found that X rays can pass through certain opaque materials. This led to their use by doctors for making photographs of bones and organs inside the body.	Gamma rays are in the frequency range just above X rays in the electromagnetic spectrum. They are also called "hard X rays" because they have greater power than X rays. It takes about 1 ft of solid lead to stop gamma rays.	Cosmic rays are at the very upper end of the electromagnetic spectrum. Thus they have the highest frequency and shortest waves of the entire spectrum. Cosmic waves, which have only recently been discovered, have even greater power than both X rays and gamma rays. They can easily pass through most materials. Cosmic rays come from outer space, and it has been found that the sun sends them out. Cosmic rays are charged particles whose speed approaches that of light.

OBJECTIVE 5 ACCOMPLISHED? FIND OUT.

1. What is a transverse wave? What is a longitudinal wave?
2. Suppose you are making a transverse wave pass along a spring. How could you increase the frequency of the wave? How could you increase the wavelength? How could you increase the amplitude?
3. Draw a model of a wave. Label the *crest, trough, wavelength* and *amplitude.*
4. What is another unit for "cycles per second"?
5. What is the cause of wave refraction? Diffraction?
6. What does polarizing do to waves?
✻ 7. Can longitudinal waves be polarized? Give a reason for your answer.
8. How does the frequency of light waves compare with that of X ray waves? How do the wavelengths of these waves compare?
9. What is the difference between light waves and the other waves in the electromagnetic spectrum?

YOUR OBJECTIVE: To find out the different ways that the speed of light has been measured; to find out how the speed of light and the theory of relativity are related; to find out what the theory of relativity says about time, length, and mass.

Before the seventeenth century, most scientists thought that light traveled from one place to another instantly. However in the latter part of the sixteenth century, Galileo Galilei (1564–1642) tried an interesting experiment to measure the speed of light (FIG. 6-1). He and a friend took lanterns to the tops of two separate

FIGURE 6-1 Galileo Galilei in his study. (*The Bettman Archive*)

hills about one mile apart. From the top of one hill, Galileo uncovered his lantern. His friend on the other hill top uncovered his lantern at the exact moment that he saw Galileo's light. Galileo measured the time it took from the moment that he uncovered his lantern to the moment he saw his friend's lantern. He thought this would be the time needed for light to travel from one hill top and back, a total of two miles.

There were two things wrong with Galileo's experiment. He did not have an accurate timing device, and the time it took the two men to remove the lantern covers was too slow. We now know that light takes only one one-hundred thousandth (1/100,000) of a second to travel a distance of two miles.

In 1675, the Danish astronomer Olaus (Ole) Roemer (1644–1710) made a fairly accurate measurement of the speed of light (FIG. 6-2). He measured the time between the moment when he saw one of Jupiter's moons disappear on one side of the planet to the moment when he saw it again on the other side. Roemer made this measurement at two different times six months apart and found a difference of about 22 minutes each time he took his measurement. Roemer thought that since the two measurements were made when the Earth was on opposite sides of its orbit around the sun, it must have taken light an extra 22 minutes to travel the extra distance of the diameter of the Earth's orbit. His result was a value of about 225 million meters per second for the speed of light.

In 1924, Albert A. Michelson (1852–1931), an American physicist, made a far more accurate measurement of the speed of light (Fig. 6-3). He set up measuring stations on two mountain tops in California. His experiment was similar to the one that Galileo had made. He placed an eight-sided rotating mirror on the top of Mount Wilson. At the top of Mount San Antonio, 22 miles away, he set up a curved mirror to reflect the light coming from the rotating mirror on Mount Wilson. The speed of the rotating mirror was carefully changed until the second reflection from that mirror appeared in the eyepiece of a viewing telescope. Knowing exactly how long it took for the mirror to rotate and the distance of 44 miles that the light traveled back and forth in that time, **Michelson calculated the speed of light to be 300 million meters per second (186,784 miles per second) in a vacuum.**

Instead of measuring the time light takes to travel a particular distance, modern researchers use measurements of the *wavelength* and *frequency of light*. These can be measured with much greater accuracy, and the simple product of one multiplied by the other gives an accurate value for the speed of light.

Distances in outer space are so great that they cannot easily be measured in the usual units that are used on the Earth's surface, such as inches, feet, yards, and miles. So space scientists have come up with a unit called the *light year*. Accordingly, **one light year is the distance light travels in one year. This is about 6 trillion (6,000,000,000,000) miles.** The star, Proxima Centauri, is about 4½ light years from the Earth. This star is the nearest star to Earth, with the exception of the sun.

FIGURE 6-2 The Danish astronomer, Olaus Roemer, looking through his homemade telescope. (*The Bettman Archive*)

FIGURE 6-3 Albert A. Michelson in his laboratory. (*The Bettman Archive*)

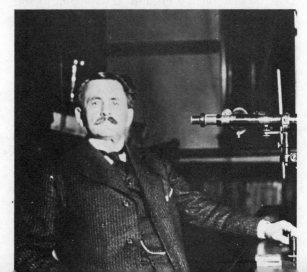

Recently, the speed of light was calculated from information gathered by two different research groups. One group worked at the National Bureau of Standards in Boulder, Colorado under the direction of Kenneth M. Evenson. This group measured the *frequency* of the waves made by a laser beam. A second group, at the Joint Institute for Laboratory Astrophysics also in Boulder, measured the *wavelength* of the same laser beam. Using these measurements of frequency and wavelength of the laser beam, they got a far more accurate measurement than had ever been obtained before. They found that light travels at a velocity of 299.7924562 million meters per second in a vacuum. This is very close to the figure that Michelson got.

Suppose two cars are moving toward each other as in Figure 6-4. If the Earth is the frame of reference, car A is going at a speed of 50 miles per hour and car B is moving at 40 miles per hour in the opposite direction. However if car A is the frame of reference, the driver in car A will see car B coming on at 90 mi/hr. This speed comes from adding the speeds of *both* cars. Then the speed of car B is 90 mi/hr *relative* to car A.

Suppose a person on the ground measures the speed of light coming from a spotlight (Fig. 6-5). The measurement will be 300 million meters per second. Suppose the speed of a rocket moving in the opposite direction is 100 million meters per second as measured by the person on the ground. If a person in the rocket measures the speed of light coming from the same spotlight, what value will be obtained? Will it be 400 million meters per second? If you think this is the answer, you are mistaken. Most scientists before this century would have given the same answer. Actually the person in the rocket ship would come up with 300 million meters per second as the speed of light from the spotlight, or exactly the same speed as measured by the person on the ground.

Many experiments have been performed to measure the speed of light in a vacuum. In some cases experimenters were moving toward the source of light. In other cases they were moving away from the source of light. There have also been cases when both the experimenter and the light source were at rest with respect to each other. In all cases, when the resulting measurements were compared, the speed of light was exactly the same.

Albert Einstein (1879–1955) was the first scientist to give an explanation for this peculiar behavior of light. **According to Einstein's *Theory of Relativity*, pub-**

FIGURE 6-4

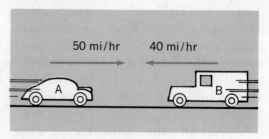

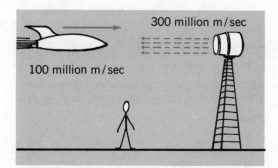

FIGURE 6-5

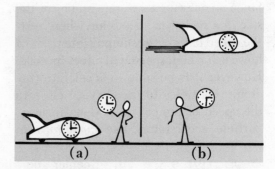

FIGURE 6-6

lished in 1905, the speed of light in a vacuum is always the same, or constant, in all frames of reference. This theory also states that nothing can travel faster than the speed of light in a vacuum.

The Theory of Relativity says nothing unusual about objects moving at speeds such as the cars in Figure 6-4. However, when objects move at very high speeds, some unusual things do happen. The Theory of Relativity explains why.

The Theory of Relativity states that time flows at different rates in frames of reference moving relative to each other. In Figure 6-6a the two clocks are in the same frame of reference, the Earth. They are at rest with respect to each other. Both clocks read the same time. Figure 6-6b shows the clock in the rocket moving at a speed very near the speed of light as measured from the Earth's reference frame. A person standing on Earth can see both of these clocks. The Theory of Relativity says that the person on the Earth sees the clock in the rocket running slower than the clock at rest on the Earth. In other words, **time measured on a fast-moving object seems to shrink.**

The Theory of Relativity also states that the length of an object in its direction of motion becomes shorter as the object moves faster. Figure 6-7a shows a figure and a rocket both at rest in an Earth frame of reference. Nothing unusual will be expected when measuring the rocket's length. Suppose now this same rocket is traveling at a speed very close to the speed of light (FIG. 6-7b). To a person on Earth, the length of this rocket will appear to be much shorter. However a person on this fast-moving rocket will notice no length difference.

What about the mass of the fast-moving rocket? Will it also change? **The Theory of Relativity states that the mass of an object will increase as its speed increases.**

FIGURE 6-7

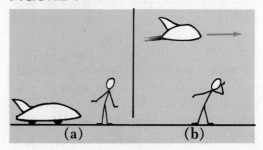

Since no one can accelerate an object the size of a rocket to speeds anywhere near the speed of light, it is impossible to show how such a high speed will affect its mass. However it is possible to accelerate tiny atomic particles to speeds very close to the speed of light. This is done in large particle accelerators like *cyclotrons* in which electromagnetic energy is used to accelerate subatomic particles, such as protons, up to 9/10 the speed of light.

Physicists have been successful in measuring a change in mass in these fast-moving particles predicted by Einstein's theory.

OBJECTIVE 6 ACCOMPLISHED? FIND OUT.

1. Roemer and Michelson used distance and time to calculate the speed of light. What measurements are used today for this calculation?
2. What is the value of the speed of light in a vacuum?
3. How does light travel? How fast does light travel?
4. What is a light year? Name two stars that are closest to Earth.
✷ 5. Calculate how long it takes light to travel from the sun to the Earth, a distance of about 150,000 million meters.
6. What does Einstein's *Theory of Relativity* state about the speed of light in a vacuum?
7. Suppose a rocket travels at a speed of 200 million meters per second. What value would a person on the ground measure for the speed of light coming from this rocket?
8. Suppose you are able to observe a rocket ship moving at a very high speed. If it begins to accelerate to a still greater speed, what would you expect to find in your measurements of its time, length and mass?

IN THIS UNIT YOU FOUND OUT

Light is a form of energy that makes it possible for you to see things. The quantum theory states that light is made up of particles (bundles) of electromagnetic waves.

A material that allows light to pass through it is called transparent. Translucent materials allow some light rays to pass through and block others. Opaque materials block light rays entirely. Absorbed light rays are rays that are stopped by a material.

Reflected light rays are rays that strike an object and bounce off. Refracted light rays are bent light rays.

A shadow forms in the area blocked off from light by an opaque object. A solar eclipse, or eclipse of the sun, occurs when the moon's shadow is seen on the Earth's surface. A total eclipse occurs when light from the sun is totally blocked from certain areas of the Earth by the moon's shadow. During a partial eclipse only a portion of the sun can be seen. When the moon enters the shadow of the Earth, a lunar eclipse, or eclipse of the moon, occurs.

Objects that are visible because they give off light are luminous. If an object becomes luminous when it is heated, it is incandescent. Objects that become luminous when struck by invisible rays are called fluorescent. Objects that do not give off their own light, but reflect it are illuminated. The amount of light given off by an object is its intensity. The intensity of a light source is measured in units called candles. The amount of light falling on an object is the illumination of the object. The unit of illuminance (illumination) is the footcandle. A light meter is used to measure illumination.

As light rays move farther away from most sources of light, they spread farther apart. The illuminance in footcandles equals the intensity of the light source in candles divided by the square of the distance in feet.

Light rays from an object are reflected, but not the object itself. A mirror reflects practically all of the light rays that strike it. When you see a likeness of an object in a mirror, it seems to be behind the mirror. This likeness is called the image. The object, itself, is called the object. The image is always located the same distance behind a plane mirror as the object is in front of it. The image of an object seen in a plane mirror is reversed. If two mirrors are placed at right angles, you will see three images. If the mirrors face each other and an object is placed between them, an infinite number of images will be reflected. When an object is placed in front of a plane mirror, the image is always the same size as the object.

When the surface of a mirror is curved inward, the mirror is concave. When the surface is curved outward, the mirror is convex. Parallel light rays that strike a concave mirror reflect toward a common point. This point, at which the rays cross, is the focus or focal point.

The principal axis is an imaginary line passing through the center of a curved mirror at right angles to the mirror. When light rays strike a concave mirror parallel to the principal axis, they reflect and cross at a focal point called the principal focal point.

An object that lies beyond the principal focal point has a real image. A real image is formed by light rays that focus at a point. An object that lies between the principal focal point and the mirror has a virtual image. A virtual image cannot be focused onto a screen like a real image can because the rays of light do not come to a focus. When an object is placed at the principal focal point, there will be no image.

When parallel rays of light strike a convex mirror, the reflected rays spread out. Because the rays that strike a convex mirror do not focus at a point, real images cannot be formed. Convex mirrors reduce the size of the image.

Light can be bent when there is a change in its speed as it passes from one material to another. This bending of light is called refraction.

A glass prism separates the colors of light because each color is refracted a different amount by the glass. A device that passes only one element and blocks all others is called a filter.

A convex lens is thicker at the center than at the edge. A concave lens is thicker at the edge than at the center. A convex lens causes light rays to focus. A concave lens causes light rays to spread out. Parallel light rays that pass through a convex lens focus at a focal point. Parallel rays that pass through a concave lens do not come to a focal point.

If an object is placed beyond the principal focal point of a convex lens, there will be a real image. If an object is at the principal focal point of a convex lens, no image is formed. If an object is placed between the principal focal point and a convex lens, a virtual image is formed.

Convex lenses are used in eyeglasses to correct hyperopic, or farsighted vision. Concave lenses are used to correct myopic, or nearsighted, vision.

A transverse wave travels at right angles to the direction of the wave's disturbance. A longitudinal wave travels in the same direction as the source of the wave's disturbance. A full wave has a crest and a trough. The amplitude, or wave height, is the perpendicular distance from a line passing through the equilibrium points to either a crest or trough.

The wavelength is the distance measured from one point on a wave to the corresponding point on the following wave. The frequency is the number of full waves — that is crest and trough — that occur in a certain length of time. The period is the time it takes to complete one full cycle.

Waves are refracted when their speed changes. Diffraction — bending and spreading out of waves — occurs when waves pass through openings smaller than one wavelength. A polarizer will allow only waves that are in the same plane as the polarizer to pass through it.

Light is a combination of electric and magnetic waves. This type of wave is called an electromagnetic wave. Michelson calculated the speed of light to be 300 million meters per second (186,784 miles per second) in a vacuum. One light year is the distance light travels in one year. This is about 6 trillion (6,000,000,000,000) miles.

According to Einstein's *Theory of Relativity*, published in 1905, the speed of light in a vacuum is always the same, or constant, in all frames of reference. This theory also states that nothing can travel faster than the speed of light. The Theory of Relativity states that time flows at different rates in frames of reference moving relative to each other — time measured on a fast-moving object seems to shrink. The Theory of Relativity also states that the length of an object in its direction of motion becomes shorter as the object moves faster. And the Theory of Relativity states that the mass of an object will increase as its speed increases.

UNIT OBJECTIVE ACCOMPLISHED? FIND OUT.

Part A Match the numbered phrases in the left-hand column with the lettered terms on the right.

1. The form of energy that can cause us to see things.
2. The "bouncing" of light rays

a. diffraction
b. Einstein
c. Huygens

from a surface.

3. The bending of light as it passes from one material into another.

4. What can be obtained from white light by using a prism.

5. Any disturbance that travels.

6. The bending and spreading out of waves as they pass through openings smaller than 1 wavelength.

7. The scientist who first made an accurate measurement of the speed of light.

8. What states that the speed of light is the same in all frames of reference?

9. The scientist who discovered that white light breaks up into various colors when passed through a prism.

10. The scientist who stated the Theory of Relativity.

d. light
e. Newton
f. photon
g. reflection
h. refraction
i. Michelson
j. Theory of Relativity
k. wave
l. visible spectrum

Part B Choose your answer carefully.

1. Most of the light passes through (a) a transparent, (b) a translucent, (c) an opaque, (d) a fluorescent object.

2. An object that becomes luminous when struck by invisible rays is (a) incandescent, (b) illuminated, (c) fluorescent, (d) transparent.

3. The amount of light given off by an object is called its (a) intensity, (b) illumination, (c) transparency, (d) fluorescence.

4. A light ray reflects because it (a) bounces off an object, (b) bends, (c) is visible, (d) produces heat.

5. The image in a plane mirror is always (a) larger than the object, (b) a real image, (c) a virtual image, (d) formed in front of the mirror.

6. Magnification is produced by a (a) convex lens, (b) concave lens, (c) prism, (d) plane mirror.

7. A real image can be formed (a) in a convex mirror, (b) by placing the object beyond the principal focal point of a concave mirror, (c) by placing the object between the principal focal point and the concave mirror, (d) from a concave lens.

8. A virtual image can be formed by placing the object   (a) on the principal focal point of a concave mirror,   (b) on a principal focal point of a convex lens,   (c) beyond the principal focal point of a convex lens,   (d) between the principal focal point and a convex lens.

9. Light refracts because of a change in its   (a) speed,   (b) illumination,   (c) frequency,   (d) color.

10. When light rays pass from air into water, they   (a) refract,   (b) reflect,   (c) get wet,   (d) change color.

11. Scientists often use the spectrum of a star to tell its   (a) size,   (b) temperature,   (c) distance from the Earth,   (d) shape.

12. The point at which light rays cross after passing through a convex lens is called a   (a) principal axis,   (b) virtual image,   (c) focal point,   (d) focal length.

13. A   (a) radio telescope,   (b) reflecting telescope,   (c) refracting telescope, (d) microscope, uses a combination of mirrors and a lens.

14. The number of waves produced in a certain period of time is called the   (a) wavelength,   (b) amplitude,   (c) frequency, (d) crest.

15. An electromagnetic wave   (a) cannot travel through a vacuum,   (b) cannot be reflected,   (c) is a longitudinal wave,   (d) is made up of an electric wave and a magnetic wave.

16. (a) Infrared waves,   (b) radio waves,   (c) ultraviolet waves,   (d) X rays,   are carriers of large amounts of heat.

17. (a) Infrared waves,   (b) ultraviolet waves,   (c) cosmic rays,   (d) radio waves,   are just above the visible part of the electromagnetic spectrum.

18. Do   (a) radio waves,   (b) X rays,   (c) gamma rays,   (d) cosmic rays,   pass most easily through matter?

19. Roemer's calculation of the speed of light made use of measurements of   (a) time and wavelength,   (b) distance and time,   (c) frequency and wavelength,   (d) frequency and time.

20. According to the Theory of Relativity,   (a) time slows down,   (b) length increases,   (c) mass decreases,   (d) length remains the same, as an object travels faster.

Part C   Think about and discuss these questions.

1. Is a convex lens or a concave lens most similar to a concave mirror? Give several reasons for your answer.

2. What is a wave polarizer?

3. According to the Theory of Relativity, how are the length and mass of an object affected by changes in speed?

ENERGY AND ITS CONSERVATION

These circus performers that you see in the unit opening photograph are using a lever to hurl one of the performers to the top position on the "pyramid." In this act the position of the fulcrum, the bar on which the lever turns, is crucial. If it is not exactly in the correct position, the performer will either not rise high enough or will rise too high to be caught by the person in the top position. This will be explained in Section 2.

1 | WORK, POWER AND ENERGY

YOUR OBJECTIVE: To understand what is meant by the terms work, power, and energy; to become familiar with units commonly used to measure work and power.

The word *work* can mean several things. Sometimes work is used to mean the method by which a person earns a living. Used in this sense, a carpenter does work while building a house. In another sense, work can mean continued activity — physical or mental — directed to some purpose. Used in this sense, you may refer to the studying of science as work. A scientist, however, is *not* using either of these meanings when using the term work.

To the scientist, work is done when a force causes an object to move. If you lift a book from the floor and set it down on a table, you have done work on the book. Work is done on an automobile when the engine causes it to move along the road. The burning fuel in a rocket is doing work when it causes a rocket to lift off the launching pad. In the examples mentioned, an object moves because of a force that acts on it. If no motion results, no work is done. You can push on a wall all day without doing any work on it. Since the wall does not move, no work is done.

You can use an equation to find out how much work is done by a force acting on an object. The equation is written

$$W = F \times d$$

The equation says that the work (W) done is equal to the product of the force (F) and the distance (d) through which the force has moved. **Work equals force times distance.**

If the force is measured in pounds, and the distance in feet, the unit for work is called the foot-pound (ft-lb). In the metric system the unit for work is the newton-meter (nt-m). The newton-meter is used when the force is measured in newtons and the distance is measured in meters. A newton-meter is usually called a joule.

When work is done to lift an object, the force needed to move it must be at least equal to its weight. For example, a force of 50 pounds is needed to lift an object weighing 50 pounds (FIG. 1-1).

SAMPLE PROBLEM: An object weighing 5 newtons is lifted from the floor to a shelf 2 meters above the floor. How much work is done on this object?

Solution: A force of 5 newtons is needed to lift the object weighing 5 newtons. This force causes the object to move a distance of 2 meters. Placing these values in the equation $W = F \times d$:

$W = 5$ newtons $\times$ 2 meters

$W = 10$ nt-m or 10 joules

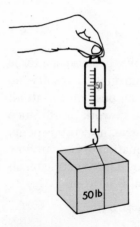

FIGURE 1-1

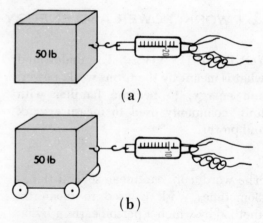

(a)

(b)

FIGURE 1-2

The force needed to move an object in a direction that is not straight upward need not be equal to the object's weight. It may be greater or less than the object's weight. However the force used to move an object must be able to overcome friction. For example, it may take only 20 pounds of force to drag a 50-pound object across the floor (FIG. 1-2a). If the 50-pound object is placed on wheels, it may take 10 pounds, or less, of force to move it (FIG. 1-2b).

Study the series of diagrams shown in Figure 1-3. Stick figure big and stick figure small are both doing work by lifting bricks from the floor onto the table. When all of the bricks have been lifted onto the table, both have done exactly the same amount of work. In what way did their work differ?

The answer is that stick figure big was able to do the work much *faster* than stick figure small. How fast work is done is what the scientist calls power. Stick fig-

FIGURE 1-3

(a)

(b)

(c)

(d)

ure big exerted more *power* while doing the same amount of work than did stick figure small. **Power is defined as the amount of work done in a certain unit of time.**

You can use the equation $P = \dfrac{W}{t}$ to calculate the amount of power used while work is being done. The equation tells you that the power (P) is equal to the work (W) divided by the period of time (t) it took to do the work.

The watt and the horsepower are two units generally used in measuring power. A watt of power is used when one joule of work is done in *one second* of time. Watts are commonly used to measure the power of radio station broadcasts, electric appliances, and light bulbs.

BE CURIOUS 1-1:

Calculate the work done on an object.

String
Brick or block of wood
Spring balances
Meter stick
Board, 2 dowels

In this investigation you will use a spring balance to lift and pull a brick or block of wood. (The extent of the weighing capacity of the spring scale you use will determine whether you use the brick or the lighter wooden block.) For each part of the investigation observe the force shown on the spring balance while the brick is moving at a *constant rate of speed.*

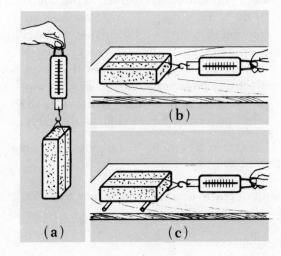

(b)

(a)   (c)

Part A

Use string to attach the spring balance to the brick as shown in Figure (a). How much work is done on the brick if it is raised to a height of ½ meter? To a height of 1 meter? To a height of 2 meters?

Part B

Pull the brick across a board as shown in Figure (b). How much work is done on the brick if it is pulled a distance of ½ meter? A distance of 1 meter? A distance of 2 meters? How do these results compare with those of Part A?

Part C

Place the brick on two dowels as shown in Figure (c). How much work is done if the brick is pulled a distance of ½ meter? A distance of 1 meter? A distance of 2 meters? How do these results compare with those in Part A and Part B? How can you explain the differences in the amounts of force used?

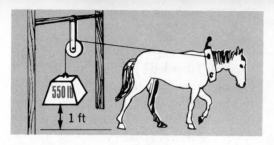

FIGURE 1-4

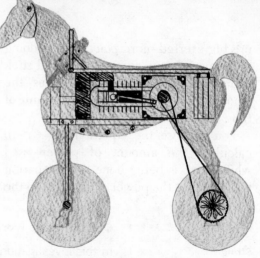

FIGURE 1-5 When engines were first built, the horse was chosen as the standard of comparison to show how much work an engine could do. The Frenchman who invented this tractor used an engine for power, but he did not give up the horse — the horse was iron.

You are probably familiar with horse-power (hp). One horsepower is defined as 550 foot-pounds of work done in 1 second. This definition was arrived at by making actual measurements. These showed that a work horse could do about 550 foot-pounds of work in one second (Fig. 1-4). The horsepower is used in rating the power of engines and electric motors (Fig. 1-5).

Energy is becoming a very familiar word. Almost every day there is news of the energy crisis or the frantic search for new sources of energy. What is energy? **Energy is sometimes defined as the ability to do work.** Energy is what enables automobiles, trains, and airplanes to carry people and goods great distances. Energy is what powers the machines that manufacture consumer products. Energy supplies light and heat. Energy provides the force necessary for motion of the body.

SAMPLE PROBLEM: A certain engine does 16,500 ft·lb of work in 10 seconds. Calculate the horsepower output of the engine.

Solution:

First find the total amount of power used by placing the values for *work* and *time* in the equation $P = \dfrac{W}{t}$

$$P = \frac{16,500 \text{ ft·lb}}{10 \text{ sec}} = 1,650 \text{ ft·lb/sec}$$

Then express the total power used in units of horsepower by dividing the total amount of power used by the amount of power that is equal to 1 hp (550 ft·lb/sec).

$$P = \frac{1,650}{550} = 3 \text{ hp}$$

1. How does a scientist define work? What two units are commonly used in measuring work? How is each defined?

2. Why is no work done on a chair when you sit on it? ✳Why is no work done on an ice skater while coasting along on frictionless ice?

3. An object weighing 7 pounds is lifted to a height of 3 feet. How much work was done on the object?

4. Suppose a box is dragged across the floor a distance of 10 meters. The box is then placed on wheels and pulled another 10 meters. During which half of the 20-meter distance was the most work done? Explain.

5. Define power. What two units are commonly used in measuring power? How is each defined?

6. Calculate the power exerted if 300 joules of work is done in a period of 5 seconds?

✳7. An elevator weighing 3 tons (6,000 lb) is moving upwards at a rate of 2 feet per second. What power is exerted in raising this elevator?

8. What is energy? How is energy related to work?

2 | MACHINES MAKE WORK EASIER

YOUR OBJECTIVE: To find out what a machine is; to identify the basic simple machines and describe how they are used to do work; to define the term *mechanical advantage;* to find out how the mechanical advantage of a machine can be calculated.

Machines are used in countless ways. **A machine is any device that is used to make work more convenient.** For example, it may be easier to push a heavy box

Device As used, something constructed for a specific purpose.

up a plank than to lift it into a truck (FIG. 2-1). So used, the plank is a machine. The engine in a truck is also a machine. It moves the truck and the box from one

FIGURE 2-1

FIGURE 2-2 This automobile jack is a lever that increases force. The handle must move through a large distance to raise the car only a few inches.

place to another. This is probably more convenient than doing the same work in some other way.

A machine cannot do work unless energy is put into it. Although the work has been made more convenient, energy is still needed to move the box up the plank — the stick figure's energy (see FIG. 2-1). The truck engine will not move the

box to another place unless it gets energy from fuel, gasoline.

Energy is put into a machine when work is done on the machine. A force moving through a distance must be supplied. The machine is then able to do work by producing *another force* that moves through a *distance* to do work. **No machine can do more work than the amount of work done upon it.** A machine does not "save" work. But, it can make work easier.

How is a machine able to make work more convenient, or easier? A machine can make work easier by increasing the force or the distance through which a force acts. If either the *force,* or the *distance* through which a force acts, is increased, the other must decrease (FIG. 2-2. Refer to unit opening photo.).

A machine can also make work more convenient by changing the direction of a force. For example, if some machines are pushed *downward* in one place, they produce an *upward* force in another place. Some machines may take a force that is moving back-and-forth and change it to a force that moves in a circle (FIG. 2-3).

FIGURE 2-3 Note the piston on this old steam locomotive. The back–and–forth motion moves a circular wheel. *(Bettman Archive)*

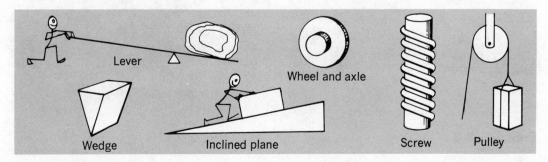

Lever

Wheel and axle

Wedge

Inclined plane

Screw

Pulley

FIGURE 2-4

There are thousands of different machines. But all of these are made by combining two or more of the six basic types of machines called *simple machines*. **The six simple machines are the lever, wheel and axle, pulley, inclined plane, wedge, and screw** (Fig. 2-4). Machines made up of only one of the six basic types are usually referred to as *simple machines*. A plank can be used as a simple machine. It is an inclined plane (see Fig. 2-1). Machines made up of a combination of *two or more* simple machines are called *compound machines*. A truck engine is an example of a compound machine.

The lever is a solid bar that rotates around a point called the fulcrum (Fig. 2-5). A force applied to the lever is called the effort. The effort causes a force to move the resistance. Levers are divided into three classes, depending upon where the effort, fulcrum, and resistance are placed with respect to one another. You will investigate a *first class*, a *second class*, and a *third class* lever system when you do *Be Curious 2-1*.

The distance from the fulcrum to the place where the effort is applied is called

the *effort arm*. The distance from the fulcrum to the point where the resistance is exerting its force is called the *resistance arm*. Like other machines, the lever can be used to increase either force or distance. It can also be used to change the direction of a force.

Suppose a certain effort is applied to a lever. The resistance that can be moved depends upon the length of the effort arm compared to the resistance arm. This relationship can be written as an equation:

$$R \times RA = E \times EA$$

This equation tells you that the product of the resistance (R) and the resistance arm (RA) is equal to the product of the effort (E) and the effort arm (EA).

FIGURE 2-5

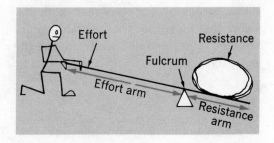

Effort

Resistance

Fulcrum

Effort arm

Resistance arm

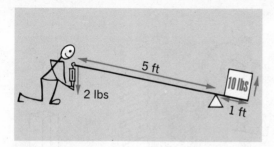

FIGURE 2-6

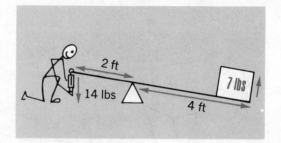

FIGURE 2-7

Study the lever shown in Figure 2-6. The product of the resistance and the resistance arm is 10 × 1. The product of the effort and the effort arm is 2 × 5. The two products are equal. The lever allows an effort of 2 pounds to lift a resistance of 10 pounds. A lever of this type *increases* the *force*. Whenever the effort arm is longer than the resistance arm, the force is increased.

Study the lever shown in Figure 2-7. The lever does not increase the force. The force is actually *decreased*. An effort of 14 pounds is needed to move a resistance of only 7 pounds. When would such a lever make work more convenient? When the effort moves down *one* unit of distance, the resistance moves up *two* units of distance. A lever of this type increases distance at the expense of force. Whenever the effort arm is shorter than the resistance arm, the distance is increased.

The mechanical advantage of a machine compares the resistance with the effort that moves it. The mechanical advantage *(M.A.)* can be found by using the equation

$$M.A. = \frac{R}{E}$$

A machine that increases the force has a mechanical advantage greater than one. For example, the mechanical advantage of the lever shown in Figure 2-6 is five. This means that every pound of effort is able to move 5 pounds of resistance. This lever increases the force. The resistance moves through a *smaller distance* than does the effort.

A machine that decreases the force has a mechanical advantage of less than one. The mechanical advantage of the lever shown in Figure 2-7 is one half. This means that every pound of effort is only able to move a half pound of resistance. On this type of lever, the resistance moves through a *greater distance* than does the effort.

A machine that does not change the force has a mechanical advantage equal to one. The mechanical advantage of the lever shown in Figure 2-8 is *one*. This means that every pound of effort is able to move one pound of resistance. On this type of lever, the resistance moves through the *same distance* as the effort. This kind of lever is used to change the *direction* of the force.

❊ The efficiency of a machine is defined as the work put out by the machine (output) divided by the work put into the machine (input). The efficiency is given as a

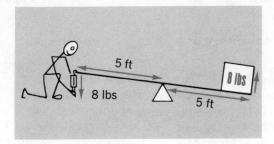

FIGURE 2-8

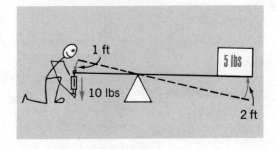

FIGURE 2-9

percent. You can state this as an equation:

$$\text{Efficiency} = \frac{\text{Work output}}{\text{Work input}} \times 100\%.$$

The work put into a machine is equal to the product of the effort and the distance through which this effort moves. Suppose an effort of 10 pounds moves a distance of 1 foot (Fig. 2-9). The work input is equal to 10 lb. × 1 ft. or 10 ft-lb. ✻ When the effort moves 1 foot, the 5 pound resistance moves 2 feet. The work output of a machine is equal to the product of the resistance and the distance through which this resistance moves. The work output of the lever is therefore 5 lb. × 2 ft. or 10 ft-lb. ✻ The lever shown in Figure 2-9 has the same work output as the work input of the machine. This machine, then, should have an efficiency of 100%. Such a machine does not actually exist. No matter how small, there is still some *friction* between the bar and the fulcrum of a real lever system. Some of the effort must then be used to overcome this friction. The rest of the effort is used to move the resistance. Only if *all* of the friction were removed could the efficiency of the lever be 100%. ✻ All machines have *more* work put into them than the work put out by them.

Therefore, all machines have efficiencies of *less* than 100%.

Another basic type of machine is called the wheel-and-axle. The larger part of the machine is called the wheel and the smaller part is the axle. The wheel-and-axle in principle resembles a first-class lever (Fig. 2-10). The fulcrum of this machine is located at the center of the wheel-and-axle system. Suppose an effort force is applied at the rim of the wheel. The radius of the wheel would then be the effort arm of a lever, and the radius of

FIGURE 2-10

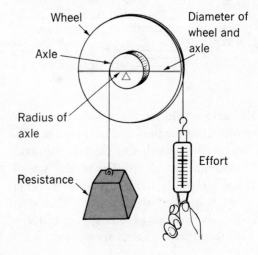

BE CURIOUS 2-1:

Meter stick
Dowel
Rubber bands, string
Spring balance
Weight

Find the mechanical advantage of each of the three classes of lever systems.

Balance a meter stick as shown in Figure (a). The rubber bands can be moved along the meter stick until it balances. Because the meter stick is balanced in this way, its own weight does not have to be accounted for in this investigation.

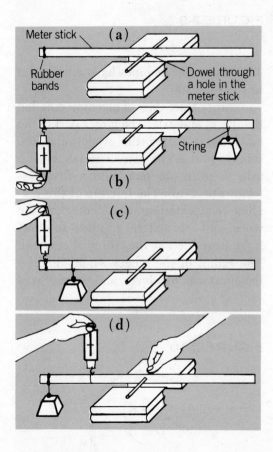

Meter stick — (a)
Rubber bands
Dowel through a hole in the meter stick
(b) String
(c)
(d)

Part A

A first class lever has the fulcrum located between the effort and the resistance. Set up a first class lever as shown in Figure (b). Move the effort and resistance to different points along the meter stick. But, always keep the fulcrum between them. Use the lengths of the effort arm and the resistance arm to predict the mechanical advantage at these points. Check your predictions from the values of the effort (read on the spring balance) and the resistance (weight). Can a first class lever have a mechanical advantage greater than one? Less than one? Equal to one?

Part B

Repeat Part A for a second class lever system. A second class lever has the resistance between the effort and the fulcrum. Set up as shown in Figure (c).

Part C

Repeat Part A using a third class lever system. A third class lever has the effort located between the resistance and the fulcrum. Set up as shown in Figure (d).

FIGURE 2-11

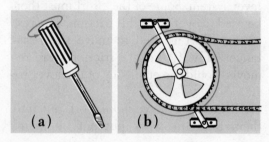

(a) (b)

the axle would be the resistance arm. A doorknob is a wheel and axle system. The knob is the wheel; the shaft is the axle.

When effort is applied to the wheel, the wheel and axle increases force. When effort is applied to the *axle*, the system increases distance at the expense of force. This results in a gain in speed (FIG. 2-11).

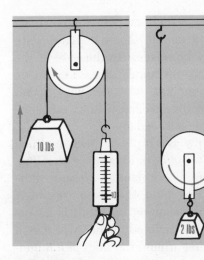

FIGURE 2-12   **FIGURE 2-13**

A pulley, like a wheel and axle, is a first class lever. You recall that a lever can be used to increase either force or distance or to change direction. A wheel and axle can increase either force or distance. So can a movable pulley. But, a fixed pulley can only be used to change the direction of a force (FIG. 2-12). It does not increase

force or distance. Suppose you wished to raise a load (resistance) using a fixed pulley. To raise a 10 lb. load, you need to pull downward with an effort of 10 pounds. If you want to raise this load up to a height of 3 feet, you must pull downward through a distance of 3 feet. How much work would you do?

A movable pulley can be used to increase force (FIG. 2-13). Two pounds of resistance can be raised with every one pound of effort. This movable pulley multiplies the effort force by 2. The pulley has a mechanical advantage of two.

Fixed and movable pulleys can be combined to form a block-and-tackle (FIG. 2-14). Each time a *movable* pulley is added, the effort force is *multiplied by 2*. Adding a *fixed* pulley does not increase the effort force. A fixed pulley serves only to change the *direction* of the force applied.

An inclined plane is a sloping surface (FIG. 2-15). Suppose two inclined planes, A and B, are used. One end of each is on

FIGURE 2-15   This shows how an inclined plane can be used to move a heavy object to a higher level. (*Editorial Photocolor Archive*)

FIGURE 2-14

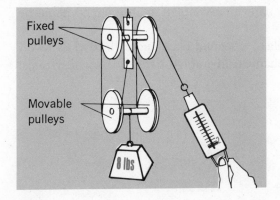

the ground and the other ends are on the same height level, say the back of a truck. Let A be longer than B, so A has a longer slope than B. It is easier to move a resistance up A because its slope is not so steep. But the resistance must be pushed a longer distance. However, even though B is shorter than A, more force must be used to push the resistance up B because its slope is steeper than that of A. Nevertheless, in both cases, the total amount of work (W) done is the same because it is the product of force (F) and distance (d), which may be written as

$$W = F \times d$$

The wedge and the screw are closely related to the inclined plane. The wedge is usually a double inclined plane, with inclines on both sides. In the case of the wedge, the resistance does not move along the incline. It is the wedge that moves and thereby forces the resistance to be lifted or forced apart.

The screw is simply a spiral inclined plane. A *straight* stairway is a machine that is an inclined plane. A *circular* stairway is a machine that is a screw. A straight road running up a hill is an example of an inclined plane, whereas a road twisting around the sides of a mountain or a ramp twisting around a building are examples of a screw.

OBJECTIVE 2
ACCOMPLISHED?
FIND OUT.

1. What is a machine? How can a machine make work easier?
2. Describe six basic simple machines. List an example of each.
3. What does the term mechanical advantage of a machine compare? ❈ Find the mechanical advantage of a machine that can move a 10-lb resistance while using a 2-lb force.
4. When does a machine have a mechanical advantage equal to one? Greater than one? Less than one?
❈ 5. What is meant by the term efficiency of a machine? What is meant by a machine being 100% efficient? Is there such a machine? Give a reason for your answer to this third question.
6. What is a fixed pulley used for? A movable pulley?
❈ 7. How can the mechanical advantage of an inclined plane be increased?
8. Describe a first class, a second class and a third class lever.
❈ Compare the mechanical advantage of the three lever systems.

YOUR OBJECTIVE: To find out more about work and energy; to learn about the different forms of energy; to understand the difference between kinetic and potential energy; to learn about the conservation of energy.

Energy was defined earlier as the ability to do work. *Work*, as you will recall, was defined, as *the product of force and distance*. Work cannot be done unless energy is present. **Work is a transfer of energy.** When work is done on an object, it receives energy. And an object is able to do work because it receives energy. Accordingly, when work is put into a machine, the machine receives energy. And a machine is able to do work only after it receives the energy.

Energy is measured in the same units as work. **One joule of energy is the amount of energy needed to do one joule of work.** In the same way, one joule of energy is transferred when one joule of work is done. Energy — like work — can also be measured in *foot-pounds* — the same as work.

There are different forms of energy. *Mechanical energy* uses the movement of matter to do work. A simple machine, such as a lever, can do work through the use of mechanical energy. *Chemical energy* is stored in fuels such as coal and gasoline. *Electrical energy* travels through wires. *Light energy* makes sight possible. Hearing is caused by *sound energy*. *Heat energy* causes objects to feel hot. *Nuclear energy* (sometimes called *atomic energy*)

is stored inside the nucleus of an atom and is released in nuclear explosions. This type of energy is used in nuclear power plants that produce electricity.

All forms of energy may be divided into two types depending upon whether the energy is *active* or *stored*. **Energy that is active, or in the process of being transferred, is called kinetic energy.** Kinetic energy is also called *energy of motion*. **Energy that is stored is called potential energy.**

Flowing water, a moving hammer and exploding gasoline, are all examples of kinetic energy (FIG. 3-1). Each is a case of kinetic energy because something is moving. And because some object is in motion, it has the ability to do work. Flowing water can do work by turning a water

FIGURE 3-1 This photo shows a jackhammer in operation. What kind of energy is being used here? *(Neal Boenzi from The New York Times)*

wheel. A moving hammer can do work by driving a nail into a piece of wood. Exploding gasoline can do work by moving the parts of an engine.

The amount of kinetic energy an object has is determined by its mass and its speed. Objects with larger masses have more kinetic energy than objects with smaller masses moving at the same speed. The kinetic energy of an object also increases with speed.

✴ The relationship of kinetic energy *(KE)* to the mass and speed of a moving object may be expressed in the form of the equation

$$KE = \frac{1}{2} mv^2$$

This equation states that the kinetic energy *(KE)* is equal to one half of the mass *(m)* of an object, times its velocity squared *(v²)*. When the mass is in kilograms (kg), and the velocity is in meters per second (m/sec), the energy has units of kilogram-meters per second squared (kg-m²/sec²), or joules.

✴ The falling brick in the energy problem has 150 joules of kinetic energy. This means that the brick would be able to do 150 joules of work. Can you think of any way that a falling brick could be used to do work?

Potential energy is stored in some form within an object. Some of the forms in which energy may be stored are: mechanical energy, chemical energy, electrical energy, nuclear energy, and so on.

An object may have potential energy because of its position. The weight in Figure 3-2, for example, has potential energy because of its raised position. If the rope that holds this weight in place is released, it will fall. As the weight falls, its potential energy will be changed into kinetic energy. The energy that was used to place the weight in its rest position was actually stored in the object. You may be certain of this because the energy used to raise the weight to a particular height, and the energy spent as it falls from this height can be measured.

Objects can have potential energy because of their condition. A spring can store energy when it is stretched. A rubber ball has potential energy when it is squeezed into a smaller volume. A flexible ruler has potential energy when it is

✴ SAMPLE PROBLEM: Suppose a brick has a mass of 3 kg. Find its kinetic energy when it is falling at a rate of 10 m/sec.

Solution: Substituting the values given for mass and velocity in the above equation,

$$KE = \frac{1}{2} (3 \text{ kg}) (10 \text{ m/sec})^2$$
$$= 150 \text{ joules}$$

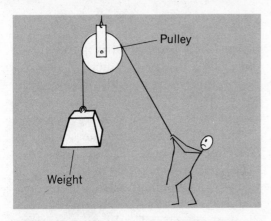

FIGURE 3-2 This stick figure is raising a weight with pulley.

FIGURE 3-3

bent. Because these objects are stretched, squeezed, or bent, they all have the ability to do work when allowed to regain their original shape.

Potential energy can be stored within atoms and molecules. If atoms or molecules move, energy is released by their motion. This stored energy is *nuclear potential energy,* or *chemical potential energy.*

Consider a ball which is thrown up into the air (FIG. 3-3). As the ball moves higher up, its speed slows down. At the same time, however, its kinetic energy is becoming smaller because its speed is steadily slowing down as it moves upward. When the ball stops for an instant at the top of its rise, it will have no kinetic energy. At that point all of its energy is potential. As the ball begins to fall, it will speed up, and its potential energy will become less and less as its height drops. However, its kinetic energy will become greater as its fall becomes faster.

The total energy can be found at any point in an object's motion by adding the **potential and kinetic energies at that point. The total energy is always the same. Therefore the energy is said to be conserved.**

Figure 3-3 shows the various positions of a ball which is thrown into the air. Its mechanical energy changes from kinetic to potential energy on the way up. Its potential energy changes back to kinetic energy on the way down. The type of energy is changing as the ball rises and falls. Is it also possible for the form of energy to change? That is, can the mechanical energy of the ball change into another form of energy?

The ball loses its mechanical energy after coming to rest on the ground. Therefore the ball's mechanical energy must change into another form of energy. Since the ball will make some sound as it strikes the ground, some of its mechanical energy must be changed into sound energy. Also the ball is slightly warmer after it hits the ground. Therefore some of its mechanical energy has also changed into heat energy.

There are many other examples of one form of energy changing into another form. Chemical energy in gasoline can be released to produce mechanical energy that can make an automobile move. Nuclear energy in a nuclear power plant changes into electrical energy. Electrical energy changes into light energy in a light bulb. A toaster changes electrical energy into heat energy.

Energy is conserved when it changes from one form of energy to another. When this happens, the total amount of energy remains the same.

If energy is conserved, how is it possible to waste it? Energy is wasted when it changes into a form that is not useful. If you touch a light bulb which has been on for a while you can find that electrical energy has been changed to heat as well as light energy. This heat energy is wasted because it is not used for anything. Therefore it is wasted energy.

The purpose of simple machines, such as levers and pulleys, is to transfer mechanical energy. But some of the mechanical energy is lost because machine parts rub against each other. The friction from rubbing parts produces sound and heat energy. This energy is wasted because it is not used.

BE CURIOUS 3-1: **See if you can recognize changes in forms of energy.**

Here are a number of devices that change one form of energy into another. List the forms of energy that are used by each device. What is the chief form of usable energy obtained from each device? Also list the unused (wasted) forms of energy from each device.

1. Complete each of the following sentences:
   Energy is the ability to do _____.
   Work is the transfer of _____.
2. What units are used to measure energy?
3. List some of the different forms of energy.
4. Compare kinetic energy and potential energy.
5. What determines the amount of kinetic energy that an object has?
* 6. Calculate the kinetic energy of a 1,500-kilogram automobile which is moving at a speed of 30 m/sec.
7. What is meant when you say that energy is conserved?

4 | SOUND ENERGY

YOUR OBJECTIVE: To find out what sound is; to learn what causes the pitch of sound; to find out how sound travels through matter; to understand the Doppler effect.

Sound is a form of energy that is caused by the vibration of matter. The vibrations can be produced in many different ways. One of the easiest ways to make a sound is to strike one object against another. A fist striking a desk will produce sound vibrations. A hammer striking a tuning fork will also produce vibrations (FIG. 4-1).

The speed at which matter vibrates is called the frequency of the sound. The lowest pitched string on a guitar vibrates slower than the highest pitched string. So

it has a lower *frequency* of vibration. Accordingly the highest pitched string vibrates more rapidly. So it has a higher frequency.

The lowest frequency sound that can be heard by a person's ear is around 20 cycles (one full swing of a vibration back and forth) per second, or 20 hertz. The highest frequency sound is around 20,000 cycles per second, or 20,000 hertz.

FIGURE 4-1

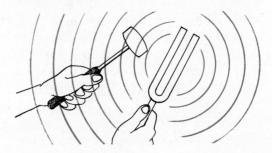

> *Vibration* The back and forth motion of matter.

FIGURE 4-2 As this young violinist moves the violin bow across a string, it makes the string vibrate and produce sound. The position of his fingers on one end of the string determines the frequency, or pitch, of the sound. *(Lincoln Center)*

The range of hearing for people is therefore from about 20 to 20,000 hertz. Some people can hear frequencies slightly above or below this range. But as people get older they have a smaller hearing range. Animals have hearing ranges different from that of people. Dogs, for example, are able to hear higher frequencies than people. A very high-pitched whistle is sometimes used by a dog owner to call the dog. The pitch of such a whistle is too high in frequency to be heard by a person.

The frequency of a sound depends on the object that vibrates. Such things as the tension (tightness), length and mass of an object determine the frequency. A greater tension on a vibrating object will produce a higher frequency. A string on a violin will produce a higher frequency as it is made tighter, or shorter — finger pressure will shorten a string (FIG. 4-2). A kettle drum will produce a higher-pitched sound when the skin of the drum is tightened. High frequencies are also made by short air columns. Therefore a trumpet will have a higher pitch than a trombone.

The greater the length of a vibrating object, the lower the frequency. The low frequencies are made by the long strings on musical instruments. Low frequencies are also made by long columns of air in such musical instruments as the tuba, trombone, or bassoon.

An object with a large mass vibrates more slowly than one with a small mass. Therefore, a larger mass tends to produce a lower frequency sound. Large bells in a church tower produce lower-pitched sounds than small sleigh bells do.

In order to hear the sound made by a vibrating drum, the sound must travel from the drum to the ear. How does sound travel from one place to another? **Sound travels in longitudinal waves.** Sound waves can carry the sound vibrations through matter. Suppose you strike a drum and cause it to vibrate. As the drum vibrates, it disturbs the air near it. As the drum head moves inward and outward, the air also moves. This movement of air, in turn, causes a sound wave which spreads out from the drum through the surrounding air. Sound waves can travel through air, or through other kinds of materials. At the Earth's surface, sound waves travel through air at about 1,100 feet per second (about 340 m/sec). But this depends on the temperature and pressure of the air. Sound waves travel faster in liquids than in gases. They travel

Observe how the frequency of vibrating matter can be changed.

Part A

Light string
Heavier string
3 test tubes

Tie one end of a piece of string to a fixed point. Pull gently on the other end. Pluck the string to make it vibrate. Pull the string tighter. What happens to the frequency (or pitch) when the string is tighter? Now change the length of the string and keep the same tightness. How does the length of the thread affect its frequency? Pluck a heavier thread and a light thread of exactly the same length with the same tightness. Which of these strings makes a higher frequency?

Part B

Fill three test tubes of the same size with water up to different levels. Blow across the top of each tube. Which tube makes the lowest sound? Notice the water level in this tube. Which tube makes the highest sound? Notice the water level in this tube. What do the results of this experiment tell you about the relation between the length of the air column in each tube, and the frequency of the sound?

still faster in solids, but they cannot travel through a vacuum.

You may have noticed that a train whistle has a higher pitch when it is approaching you. And the pitch then seems to become lower as soon as it passes you. **A change in pitch because of the motion of a source of sound is called the Doppler effect.**

Suppose you are making a rope vibrate by moving your hand rapidly up and down. The vibration will travel along the rope as waves. The frequency of the vibrations or waves can be kept constant if your hand moves up and down steadily (FIG. 4-3a). Suppose you walk toward the wall without changing the frequency of your up-and-down hand motion (FIG. 4-3b). As you do so the waves on the rope will become closer to each other. This is the same as if you stood still but increased

the frequency of your hand's motion. Now suppose you back away from the wall, all the while keeping the same frequency of your hand's motion (FIG. 4-3c). You should see that the waves are becoming

FIGURE 4-3

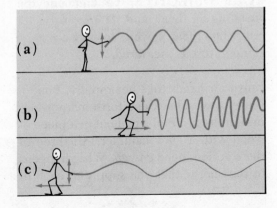

(a)

(b)

(c)

farther apart. This is the same as if you stood in the same place but slowed down the frequency of your hand's motion. Approaching sources of sound seem to have a higher frequency than the actual frequency produced. The frequency seems lower for sources that are moving away.

The Doppler effect can also be seen with other types of waves. Light waves reaching the Earth from distant stars show a Doppler effect because these stars are in motion toward or away from the Earth.

OBJECTIVE 4
ACCOMPLISHED?
FIND OUT.

1. What is sound?
2. What is meant by frequency?
3. What is the approximate hearing range of the human ear?
4. What happens to the frequency of a vibrating string if it is made tighter? Longer? Heavier?
5. A xylophone has short metal bars at one end and longer bars at the other end. Which of the bars will give the highest pitch? Give a reason for your answer.
6. Why is it impossible for sound to travel through a vacuum?
7. What is the Doppler effect in sound?
✳ 8. Suppose the stick person in Figure 4-3b stood still, and the wall moved toward the hand making the waves on the rope. What would happen to the waves along the rope?

5 | HEAT AND TEMPERATURE

YOUR OBJECTIVE: To find out the meaning of heat and temperature; to learn how heat travels; to find out how temperature is measured.

Rub your hands together rapidly. Bend a piece of wire back and forth many times until it breaks. Pull a nail out of a piece of wood with a claw hammer. All three of these actions need *energy*. What happens to the energy that you supply to each of

these actions? You will notice that your hands become warm after rubbing them together. If you touch the wire that you bent back and forth at the point where it broke, you will feel that it is also warm. And you will find that the nail is quite warm at the end that was in the wood. In each of these actions mechanical energy was turned into heat. **The form of energy that causes the temperature of an object to change is called heat energy.**

Heat, like other forms of energy, can be measured in units of joules (named for James Prescott Joule) or foot-pounds. Another unit used for measuring amounts of heat energy is the calorie.

You have probably used calorie units when talking about the energy stored in foods. One calorie is the amount of heat energy needed to increase the temperature of one gram of water by one Celsius degree (discussed further on in this section).

The British thermal unit (BTU) is still another unit of heat measurement. You are perhaps familiar with the BTU in connection with air conditioners. It is also used in connection with furnace fuels.

Heat travels through solid materials by conduction. In solid materials the atoms are not free to flow about. Heat energy must be passed from one atom to the next. To see this, suppose one end of a metal spoon is held in a candle flame (Fig. 5-1). In a short while, the opposite end will begin to feel warm in your hand. This end becomes warm because of *conduction*.

The atoms nearest the flame receive heat energy first. Then it passes on to neighboring atoms. Those atoms in turn, pass the energy on to other neighboring atoms through the spoon until the end in your hand is reached. The heat energy seems to flow through the spoon. The atoms passing this energy do not move up the spoon. They remain in the same place, but vibrate more rapidly.

Some substances allow heat to be conducted through them more easily than others. These substances are called *heat conductors*. Metals such as aluminum, copper and iron are examples of good heat conductors. Other substances allow very little heat to be conducted through them. These are called *heat insulators*. Asbestos, wood, and plastics are examples of good insulators.

In liquids and gases, the atoms and molecules move about more freely than in solids. Instead of passing heat along from one particle to the next, the atoms and molecules carry the energy along from one place to another. **The movement of**

FIGURE 5-1

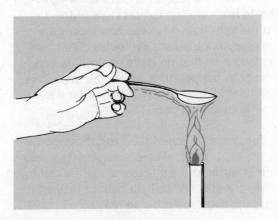

FIGURE 5-2 As the water becomes hotter near the bottom of the beaker, the coloring matter flows upward, as shown by the arrows.

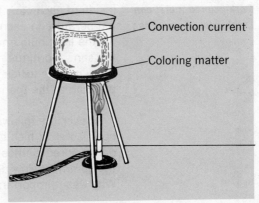

Convection current

Coloring matter

heat caused by the flow of atoms and molecules is called convection. Figure 5-2 shows water being heated in a beaker. The water nearest the flame becomes warmer. The warmer water then rises and nearby cooler water takes its place. This cooler water is heated, in turn, by the flame. It then rises in the same way. As this continues, heat energy spreads throughout the water.

Heat energy can also travel by *radiation* (Fig. 5-3). **The movement of heat energy in the form of invisible waves is called heat radiation.** These waves can pass through such materials as air and glass. The waves travel even better through a vacuum. Such waves are called *infrared waves.* If an object absorbs more heat radiation than it gives off, it gets warmer. If it absorbs less heat than it gives off, it gets cooler. An object that absorbs the same amount of heat as it gives off stays at the same temperature.

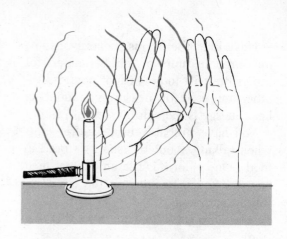

FIGURE 5-3

Chemical energy stored in fuels can be changed into heat energy when fuels are burned. Electrical energy is changed into heat energy by an electric stove or a hair dryer. Much heat energy can be obtained from the nuclear energy stored inside the atom. In the same way, heat energy can be changed into other forms of energy.

BE CURIOUS 5-1:   **Observe the effect of color on heat radiation.**

Light bulb and socket
2 thermometers
Dark-colored metal can
Light-colored metal can

Place light and dark metal cans at equal distances from a light bulb. Use a thermometer to measure the temperature of the air inside of each can. Turn the light bulb on. Read and record the temperature in each can every minute for 15 minutes. Then turn off the light bulb and remove it so that no more heat will fall on the cans. Continue reading the temperature on the thermometers every minute for an additional 15 minutes. *Note:* at no time should the bulb of the thermometer touch the can.

Answer the following questions: Which of the two cans absorbed heat faster? Which gave off heat faster? Why do people often wear dark-colored clothes during winter and light-colored clothes during the summer? How did heat travel from the light bulb to the cans? Why was it important that the two cans differ in only color?

For example, steam engines and turbines use heat energy to produce mechanical energy.

You have learned earlier that heat energy causes temperature to change. But what is meant by temperature? **Temperature is the measure of the "hotness" or "coldness" of a substance.** When an object or substance gets hotter, its atoms move around faster. Therefore, atoms have more kinetic energy when they get hotter. When a substance gets colder, its atoms slow down. These atoms then lose some of their kinetic energy. Therefore, the "hotness" or "coldness" of an object or substance is related to the kinetic energy of its atoms.

Temperature is usually measured by a thermometer. One scale used to measure temperature is called the *Fahrenheit* scale. This scale is commonly used in weather reports. It was named after its inventor, the German physicist, Gabriel Fahrenheit (1686–1736). On the Fahrenheit temperature scale, water freezes at 32°F and boils at 212°F. The average room temperature on this scale is about 70°F. In scientific work, however, the *Celsius*, or *centigrade*, scale is used. This scale was developed by Anders Celsius (1701–1744). On the Celsius temperature scale, water freezes at 0°C and boils at 100°C. A comparison between the Fahrenheit and Celsius scales is shown in Figure 5-4.

✱ Since the temperature of an object or substance cools as its atoms slow down, is there a limit as to how cold a substance or object can get? You would think that the coldest possible temperature would be when atoms stop moving completely. Ex-

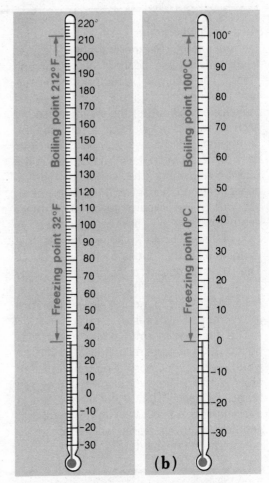

FIGURE 5-4 (a) Fahrenheit thermometer and (b) Celsius thermometer.

periments show that this happens very close to 460°F or 273°C below 0° on either scale. There is a third temperature scale called the *Kelvin*, or *absolute* scale. This scale is also used in scientific work. On the Kelvin temperature scale, the lowest possible temperature of a substance is the 0°K or absolute zero. This is the same as 460°F and 273°C below zero on the Fahrenheit and Celsius scales.

Heat from inside the Earth is called *geothermal energy*. Geothermal energy comes from underground pools of hot water and steam. The steam is brought up to the surface and used to run the machinery that makes electricity. After the steam is used, it is changed into hot water and returned underground. The hot water then is changed back to steam when it again reaches the hot rock underground. The use of geothermal energy does no damage to the environment.

Some energy experts favor using more geothermal energy. They point out that this type of energy could make enough electricity for the Earth's needs by the year 2000. The largest amounts of geothermal energy are found in areas of *hot springs* and *geyser fields*. These are found in Yellowstone Park in Wyoming, New Zealand, and Iceland.

The first geothermal power plant was built in 1904 at Larderello, Italy. It is still the world's largest geothermal power plant. Another geothermal power plant, north of San Francisco, called The Geysers, has been in operation since 1960. Why do you think that geothermal power plants will be most important in the near future?

OBJECTIVE 5
ACCOMPLISHED?
FIND OUT.

1. Define heat.
2. What units can be used to measure amounts of heat energy?
3. How does heat travel through substance such as iron and copper? How does heat travel from sun to Earth?
4. What is a heat conductor? What is a heat insulator?
5. What happens to the temperature of objects that give off more heat by radiation than they absorb?
6. How is the temperature of a substance related to the motion of its atoms?
7. What are the freezing and boiling points of water on the Celsius scale? On the Fahrenheit scale?
✱ 8. What is meant by absolute zero?

6 | ELECTRICAL ENERGY

YOUR OBJECTIVE: To find out about static electricity and electron flow, voltage, and resistance; to understand how batteries, generators, and motors work; to learn how electrons flow in series and parallel circuits; to find out how some electronic devices are used.

Electrical energy is used for many things. You are probably familiar with electric lights and small appliances that use electricity, such as radios, televisions, vacuum cleaners, toasters, steam irons and hair dryers.

Electrical energy is called electricity. It comes from negatively charged particles in atoms called electrons. Static electricity is an isolated charge (negative) which may either remain motionless on an object or may flow for a brief time from one object to another. Static electricity flows from one object to another as a result of contact or friction.

Since all matter is made up of atoms that have electrons orbiting the nucleus, all matter has some electrical property. When you walk across a wool rug and then touch a metal doorknob you will probably get a "shock." That is, a tiny spark will occur when you touch the doorknob. This is static electricity. Whenever you see such a spark there is electron movement. This is called an *arc*.

Particle Bits of matter that are so small they cannot be seen.

However, if you touch a glass doorknob there will be no spark. Why?

When you walk across a wool rug, your feet "rub" electrons off the rug and these collect on your body. You could say that your body is *negatively charged*, because it has an abundance of electrons, and that the doorknob is *positively charged* because it has a shortage of electrons compared with your body after walking across the wool rug. Electrons will flow from a negatively charged point to a positively charged point.

Since electrons moved from your negatively charged body into the positively charged doorknob, the doorknob is said to be a *conductor* because it "draws" the electrons and causes them to flow from your body. A glass doorknob on the other hand, will not conduct electrons. You can see this because if you touch a glass doorknob after walking across a wool rug, there will be no spark or shock. Therefore glass is said to be an *insulator*, because it does not conduct electrons. Some good electrical conductors are copper, silver and iron. Some insulators are glass, rubber, wood and plastic.

When you see lightning during a thunderstorm, you are seeing a huge spark. This is because electrons in a negatively charged cloud are "jumping" across or *arcing* across air to a positively charged cloud or a neutral earth. The arcing electrons cause the air around the arc to expand quickly. This action causes thunder.

What is usually considered to be an electric current is a continuous flow of electrons through a conductor, which is

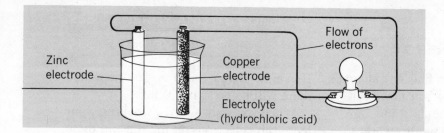

FIGURE 6-1   Wet cell.

Zinc electrode

Copper electrode

Flow of electrons

Electrolyte (hydrochloric acid)

usually a metal wire. But in order for current to flow, a circuit, or closed loop of wire, is needed.

A force, called an electromotive force (EMF), or voltage, is needed to make electrons flow through a wire. The two most common sources of voltage are batteries and generators.

A battery consists of one or more wet or dry electric cells. A wet cell is a container filled with a liquid called an electrolyte (Fig. 6-1). Two metal plates are also contained in a wet cell. One plate may be copper, and the other zinc. These materials contain electrons that can be made to flow as electric current. The copper and zinc plates are called *electrodes*. The electrolyte may be hydrochloric acid.

One electrode will contain more electrons than the other. This electrode will be negatively charged. The electrode that contains the fewest electrons will be positively charged. For this reason there will be a *potential difference* between the two electrodes. This potential difference will set up a voltage between the electrodes. And this voltage, in turn, will cause electrons to flow from the electrode with the most electrons to the electrode with the fewest.

A battery of wet cells, or storage battery, consists of one or more wet cells connected in series into a circuit. This means that the cells are connected, one after the other, negative end to positive end. The battery used in an automobile contains several wet cells connected in this way.

A typical dry cell is used in a flashlight battery (Fig. 6-2). A dry cell is usually a zinc container filled with a moist chemical paste, usually manganese dioxide. A carbon rod runs through the center of the cell through the paste. The zinc in the container wall serves as the negative electrode and the carbon rod is the positive electrode. Again, a potential is set up between these electrodes that causes electrons to flow. In flashlights and portable radios, one or more dry cells, usually just called batteries, are connected in series to provide the proper amount of voltage needed for operating these appliances.

FIGURE 6-2   Dry cell.

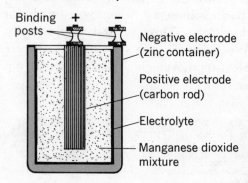

Binding posts

+   −

Negative electrode (zinc container)

Positive electrode (carbon rod)

Electrolyte

Manganese dioxide mixture

Find out what happens when a piece of wire is placed in a magnetic field.

Part A

1 piece of straight
   copper wire
1 galvanometer
1 horseshoe magnet
2 alligator clips
1 piece of coiled copper
   wire
1 bar magnet

A galvanometer is used to measure small amounts of electron flow in a wire. Connect each end of the straight copper wire to a terminal of the galvanometer as shown in Figure (a). Pass the wire between the poles of the horseshoe magnet. What happens to the pointer on the galvanometer? When the magnet is removed, does the galvanometer pointer move? Now move the wire rapidly back and forth between the magnet's poles. What happens to the pointer now?

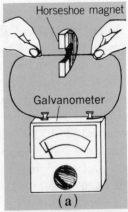

Horseshoe magnet

Galvanometer

(a)

Part B

Now connect the piece of coiled wire to the galvanometer. Pass the bar magnet through the coil once as shown in Figure (b). What happens on the galvanometer? Now move the magnet back and forth rapidly through the coil. What does the pointer on the galvanometer do? How does the movement of the galvanometer now compare with Part A when you moved the straight wire rapidly between the poles of the horseshoe magnet?

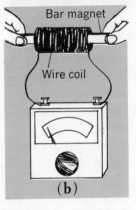

Bar magnet

Wire coil

(b)

Electrons flow through a circuit from negative to positive. One-directional flow of electrons is called direct current (dc).

In *Be curious 6-1* you saw the pointer on the galvanometer deflect when a piece of wire was placed between the poles of the magnet. What you did was to place a *circuit* of wire in the magnetic field between the magnet's poles. (The circuit is completed, or closed, in the galvanometer.) The deflection of the galvanometer told you that there was some electron flow in the wire. But you noticed that

when you stopped moving the wire between magnet's poles, the pointer returned to the zero position, where it was when you began your experiment. Then when you moved the wire rapidly back and forth between the magnet's poles, the pointer moved back and forth. In fact, **the strength of the current in a wire moving in a magnetic field depends on the speed of the wire's motion in the field and on the strength of the magnet.** As you shall see, this is the principle on which an electric generator works.

The galvanometer you used in your experiment contained a magnet and a coil of wire placed in the magnet's field. When you passed the copper wire between the poles of the horseshoe magnet, you *induced* an electron flow in the wire. These electrons flowed into the galvanometer through its wire coil. This, in turn, caused the magnet to turn. The pointer is connected to the magnet. Some galvanometers have the pointer connected to the coil which moves when electrons flow through the magnet (FIG. 6-3). In any case the working principle is the same — electron flow is induced in a wire moving in a magnetic field.

When electrons flow through a straight piece of wire, a certain magnetic field will build up around the wire. But when the wire is coiled, like thread on a spool, the magnetic field will become much stronger. Michael Faraday (1791–1867) a British physicist and chemist connected a battery to a coil through a switch. He placed a second coil connected to a galvanometer near the first coil, but not touching it. Then he closed the switch connected to the first coil. This caused electron flow from the battery through the first coil. He noticed that the pointer on the galvanometer, which was connected to the second coil, deflected. The result of this experiment is that **electron flow in one coil of wire can be induced into a nearby coil of wire because of a magnetic field that builds up between the coils. This is called electromagnetic induction.** This is the principle that led to the development of a device called a *transformer,* which can increase or reduce electron flow in the second coil. Transformers are used in power plants to increase voltage. The principle of electromagnetic induction was also discovered by the American physicist, Joseph Henry (1797–1878).

An electric generator changes mechanical energy into electrical energy. It consists of a coil of wire, called an armature, that is made to turn between the poles of a magnet (FIG. 6-4). As the armature turns, an electric field is set up that causes electrons to flow in the coil of the armature. However the armature passes *across* magnetic lines of force between the poles of the magnet. Therefore as the

FIGURE 6-3 Moving-coil galvanometer.

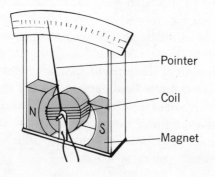

Pointer

Coil

Magnet

FIGURE 6-4 Hand generator.

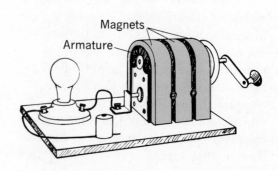

Magnets

Armature

coil turns in one complete cycle, or circle, the amount of voltage in the circuit will vary. When the voltage goes from zero to its positive peak, electrons flow through the armature in one direction, but when the voltage returns through its negative peak, the electrons will flow in the reverse direction. **Current that flows in one direction and then in another is called alternating current (ac).**

The armature can be turned at different speeds. The rate at which the armature turns determines the frequency of the alternating current.

The voltage that supplies electricity to most buildings is alternating so it supplies alternating current. This voltage comes from huge generators in electric power plants. The most common ac frequency is 60 cycles per second, or 60 hertz.

Voltage is measured in units called volts. A voltmeter is used to measure voltage (FIG. 6-5). Two wires called leads are attached to a voltmeter. Direct voltage must be measured by connecting one lead to the positive terminal, or point, on the circuit being measured, and the other lead to the negative point. For example, if you want to measure the voltage of a battery, you must connect the positive lead from the voltmeter to the positive terminal on the battery and the negative lead to the negative terminal on the battery. The positive lead is usually red and the negative lead is black. A voltmeter is connected *across* a circuit so it is parallel to the source of voltage. When alternating voltage is measured it makes no difference which lead is used because the meter will measure the average voltage. This is as if it were direct voltage.

FIGURE 6-5 This multi–function meter contains ac and dc voltmeters, and an ohmmeter. The dial at the left selects the meter. The dial at the right selects the range of voltage. Wire leads may be connected to the three terminals (at bottom). The label, COM, stands for "common" or ground, and is used on the dc meter. *(Hewlett-Packard)*

Electron flow (or current) is measured in units called amperes. The unit ampere is named for French physicist, André Ampère (1775–1836). Electron flow is measured by an instrument called an ammeter. To measure electric current, an ammeter must be placed *in series* with the voltage source. In a dc circuit, powered by a battery, the negative lead of the meter is attached to the wire in the circuit toward the negative terminal of the battery, and the positive lead is attached toward the positive terminal of the battery.

An electric motor changes electrical energy into mechanical energy. This is the reverse of what happens in a generator. A *field coil* is wrapped around a magnet, and dc voltage from a battery is applied to this coil and to the coil of the armature (Fig. 6-6). This dc voltage causes the armature coil to become "magnetized" so that it has a north pole and a south pole like the magnet. As *like* poles of the magnet and armature are repelled, the armature is moved. To keep the armature moving, a *commutator* and *brushes* are used to change the direction of current flow. This reverses the poles of the armature to keep it in motion. The moving armature can be used to drive a mechanical device. Hence this is a motor. It is a dc motor because dc voltage is supplied to operate it.

Some type of materials help electrons to flow more easily through an electric circuit. Such materials are called conductors. Other materials oppose or slow down the flow of electrons. These materials provide *resistance* to a circuit. All materials through which electrons flow have some resistance. But if there is almost no resistance in a circuit, there is a very large amount of current. Such a circuit is called a short circuit. Short circuits are very dangerous because of the large amount of current. They often cause fires. The greatest resistance occurs when electron flow is completely blocked. This occurs when a circuit is *open,* or disconnected. Materials like rubber and glass completely block the flow of electrons because they are *insulators.*

An element that provides electrical resistance in a circuit is called a resistor. All circuits require resistors to prevent short circuits. A familiar kind of resistor is the electric light bulb. You know that if you touch a light bulb that has been on for a while, it will be very hot. So one characteristic of a resistor is heat, or heat energy. But not all resistors change electrical energy into heat energy as do light bulbs and electric irons, toasters, stoves and hair dryers. Radios, televisions, and electric fans give off some heat, but not as much as the applicances just named. However, *all* appliances are resistors. **The unit of resistance is the ohm.** The Greek letter *omega,* Ω, is used as an abbreviation for *ohm* in circuit drawings.

The amount of resistance offered to a circuit by a resistor depends on the material that the resistor is made of as well as its size. For example, carbon, lead, and German silver offer more resistance than copper, silver, and nichrome — nichrome wire has a resistance that is 66 times

German silver   A white metal made of copper, nickel and zinc.

FIGURE 6-6

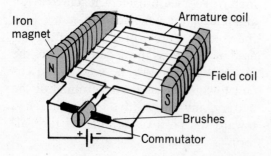

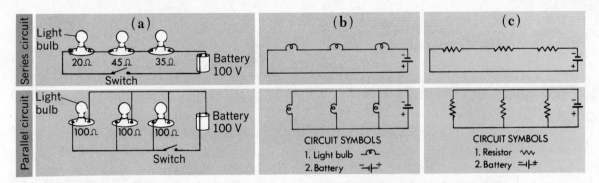

FIGURE 6-7 (a) Series and parallel circuits using light bulbs for resistances; (b) using circuit symbols for light bulbs and battery; (c) using resistor symbols.

greater than that of copper. Since high resistance produces more heat, nichrome is often used in hot plates and toasters. A thin wire will have more resistance than a thick one of the same kind of metal. Also the longer a piece of wire is, the greater its resistance is.

Since a switch is a device that opens and closes a circuit, when it is closed it acts as a perfect conductor, because it allows electrons to flow through it. However when it is open, it acts as an *infinite*, or perfect resistance, because it prevents any electrons from flowing through it.

In 1827, a German physicist, Georg Ohm (1787–1854) showed how current, voltage and resistance are related. He expressed this relationship which is now called *Ohm's Law*.

Ohm's law says that the current depends directly on voltage and indirectly on resistance. This means that current will become larger as the voltage gets larger, and that the current will become larger as the resistance gets small, and conversely. If I is the current, V is the voltage, and R is the resistance, then, Ohm's law can be written as $I = V/R$. The other forms of Ohm's law are $V = IR$ and $R = V/I$.

Resistances can be connected in series or in parallel in a circuit (FIG. 6-7). **Resistors are in series if they are placed in a**

Conversely In the reverse order.

SAMPLE PROBLEM: A light bulb has a resistance of about 200 ohms. What is the current flowing through this bulb if the voltage is 110 volts?

Solution:

Using Ohm's law and the proper units for current, voltage and resistance,

$$I = \frac{110 \text{ volts}}{200 \text{ ohms}} = 0.55 \text{ ampere}$$

circuit one after the other. When light bulbs are connected in series, and one goes out, the circuit will act like a broken circuit and all of the bulbs will go out. This is because the current is the same everywhere in a series circuit. As more and more bulbs are added in series, they will all become dimmer. This is because each bulb is a resistor, and as more resistors are added to a circuit in series, electron flow is retarded by each resistor. The total resistance in a series circuit can be found by adding each of the resistances. For example suppose there are three light bulbs in a circuit (see Fig. 6-7a). One is 20 ohms, the next is 45 ohms, and the last is 30 ohms. Then the total resistance in the circuit is 20 + 45 + 35 = 100 ohms.

Resistance can be measured with an ohmmeter (see Fig. 6-5). The leads of an ohmmeter are connected across the resistor or resistance to be measured. The reading will be in ohms.

If light bulbs are connected in parallel and one is removed the other bulbs will remain lighted. This is because the *circuit* is still connected through parallel paths. You have seen a Christmas tree with one or two bulbs out and the rest of the bulbs still on. This is because these small bulbs are all connected in parallel. If they were connected in series they would all go out because the circuit would be broken. Furthermore you couldn't tell which bulb was bad without testing each one separately. This is why electrical appliances in a home are connected in parallel. The operation of one appliance does not affect the others. If they were in series, turning off or removing one appliance would cause the others to stop working.

The electrical power used in operating most appliances is measured in units called watts. One watt is the amount of power used when one joule of work is done in one second. A 60 watt light bulb then uses 60 joules of energy every second when it is on.

Electricity is purchased according to the amount of work it does. Therefore,

BE CURIOUS 6-2: **Find out how resistances in series compare to those in parallel.**

Part A

Battery
3 flashlight bulbs
3 bulb sockets
Wire

Connect one electric bulb to a cell or battery. Observe the brightness of the bulb. Connect a second bulb in series with the first. Is the other bulb affected in any way? Take one of the bulbs out of its socket. Does this affect the other bulb? Connect a third bulb in series. What are the results?

Part B

Repeat Part A, this time connecting all bulbs in parallel. How do the results differ? How do you know that the light bulbs in your home are connected in parallel?

FIGURE 6-8 (a) Here you can see two different size vacuum tubes with glass envelopes and the tiny point–contact transistor (center). (b) Here are two vacuum-sealed transistors. They can resist high–energy radiation in outer space. This type of transistor is used in Telestar II. (c) Here is a "printed circuit" board. The round metal disks are transistors. (*Western Electric; Hewlett-Packard*)

you pay directly for the amount of energy that is used in an electrical appliance.

Electrical energy is usually purchased in units of kilowatt-hours. A kilowatt-hour is equal to 1,000 watts of power used for one hour.

Electronics deals with the flow of electrons through vacuum tubes or transistors (FIG. 6-8). Just as a faucet is used to control the amount of water flowing in a pipe, vacuum tubes and transistors control the amount of electron flow in a circuit.

Vacuum tubes are usually made of a glass or metal shell called the *envelope*. All of the air is removed from the envelope. This is why they are called *vacuum* tubes. Glass vacuum tubes look something like light bulbs. Small changes in voltage applied to a vacuum tube can cause large changes in the current flowing through the tube. This is called *amplification*. Therefore vacuum tubes can amplify small signals, or sound waves, like voice and music signals that come from a mi-

crophone. The microphone changes sound waves moving through air to electrical energy. This electrical energy is fed to amplification that contains vacuum tubes or transistors.

Transistors work in a similar way to vacuum tubes but they are made from materials called *semiconductors*. The resistance of a semiconductor lies somewhere between that of a conductor and an insulator. Again when small changes of voltage are applied to a transistor, large changes occur in the current flowing through it. Since vacuum tubes and transistors amplify electrical signals, they are often called amplifiers.

Radio sets receive very weak electrical signals from radio waves in the air. These weak signals are amplified by vacuum tubes or transistors in a radio. Strong electrical signals result. These signals are applied to speakers in the radio. The speakers change this electrical energy back into sound waves, and they can be heard.

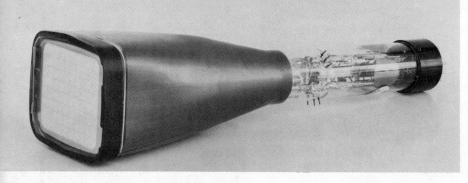

FIGURE 6-9 This is a typical cathode–ray tube. *(Hewlett-Packard)*

Television sets also receive weak electrical signals from the air. Some of these signals are amplified and produce sound. Other signals are amplified and produce the picture.

The cathode-ray tube is a special type of vacuum tube (Fig. 6-9). The picture on a television set is seen on the screen of a cathode-ray tube. The inside surface of the screen is coated with a *fluorescent* material. This coating gives off light when struck by a beam of electrons. The number of electrons in that beam determines the brightness of the light given off. A large number of electrons cause brightness on the screen. Dark areas are made by cutting off electrons in the beam.

Cathode-ray tubes are also used in other types of electronic equipment. A radar set uses this kind of tube to show the location of ships or airplanes. It can also show the location of approaching storms.

OBJECTIVE 6 ACCOMPLISHED? FIND OUT.

1. What is the difference between static electricity and electron flow?

2. What is electric current, voltage, and resistance?

3. What is the difference between ac and dc voltage or currents? Which type is produced by a chemical cell? Which is found in the wiring in your home?

* 4. A light bulb with a resistance of 50 ohms is connected to a battery. Find the current through the bulb if the battery produces 25 volts.

5. Draw a picture of three household electrical applicances connected in series to a dry cell. Draw a picture of these items connected in parallel.

6. What unit is commonly used for measuring voltage? Current? Resistance? Electrical power? What kind of meter is used to measure volts? Amperes? Ohms?

* 7 What causes light to appear on the screen of a cathode-ray tube?

7 | ENERGY FROM THE NUCLEUS

YOUR OBJECTIVE: To find out about radioactivity; to learn about radioactive decay; to find out how atomic fission and fusion produce energy.

In 1896 the French scientist, Antoine Henri Becquerel (1852–1908) made one of the most exciting and important discoveries of the century. He had stored some uranium ore in a drawer which also contained photographic film. The film was wrapped in heavy black paper to protect it from light. Later on, Becquerel found that the film in this drawer looked as if it had been exposed to light, although it had not been removed from the black paper wrapping. Finally, it occurred to him that perhaps the uranium ore gave off some invisible form of radiation which passed through the protective black paper to expose the film. A few experiments proved his suspicion to be correct. Becquerel had discovered radioactivity.

Radioactivity is the release of invisible rays from the nucleus of an atom. Immediately after Becquerel's discovery, other scientists began to study radioactivity (FIG. 7-1). They found that other substances, such as thorium and radium, also produced invisible rays. For this reason, these substances are called radioactive elements.

Further experiments showed that there were three kinds of invisible rays. The first kind was made of fast-moving electrons called beta rays (or beta particles). The second type of ray called *alpha rays* are made up of much heavier particles

FIGURE 7-1 Marie Sklodowska Curie and her husband Pierre discovered polonium and radium. For their work on radioactivity, they shared with Becquerel the 1903 Nobel prize in physics. *(Bettman Archives)*

(alpha particles) than the electrons. The third kind of ray, which behaves more like X ray waves are known as *gamma rays*. Further experiment showed that gamma rays were very similar to X rays. They easily passed through opaque materials that light rays could not enter. In some cases they pass through substances more easily than X rays. The radioactivity discovered by Becquerel is called *natural radioactivity*. As time passes the nucleus of an atom will grow smaller and a different kind of atom will result. This is called *radioactive decay*. For example, when a certain type of uranium atom decays, a thorium atom will result. An alpha particle will be released when this happens. Therefore there will be an alpha ray. Thorium is also radioactive. After a period of time, an atom of protoactinium will result. A beta particle will then be released. So there will be a beta ray.

The protoactinium will be radioactive. This atom will also decay into a still smaller nucleus. This will continue until a nucleus of lead results. Then all radioactivity will stop. Therefore the lead nucleus is said to be *stable*. So no further decay will take place. **Radioactive decay is measured by a unit called half-life. The half-life is the time it takes for half of the atoms of a radioactive sample to decay.** Suppose you have 16 grams of a radioactive material whose half-life is known to be 1 hour. At the end of 1 hour, only 8 grams will be changed into a different substance by radioactive decay.

The value for the half-life is different for different substances. The half-life of uranium is about 4½ billion years. It takes only 24 days for half of the thorium atoms to decay into protoactinium. Some radioactive substances have a half-life of only a small fraction of a second.

Although the rays from a radioactive source are invisible, they can easily be detected. Becquerel first discovered radiation by using photographic film. Today scientists still use various types of film to do this. Film is sent up in balloons and rockets to record radiation at great heights above the Earth. Laboratory workers in areas of radiation must wear *film badges* to record the amount of radiation that reaches them.

Certain gas-filled tubes are used to detect radiation. When radiation passes through a tube it causes a short "burst" of electricity. Each burst can be heard as a clicking sound on a *Geiger counter*. Geiger counters can tell how much radiation there is in a certain place.

The amount of radiation coming from rocks helps scientists determine a rock's age. This is known as *radioactive dating*. It enables scientists to tell how many years ago prehistoric plants and animals lived on the Earth.

In 1919, the English scientist, Ernest Rutherford (1871–1937) forced radiation from a nucleus by splitting it. He aimed alpha particles from radium at some nitrogen gas. The alpha particles from the radioactive radium were used as "atomic bullets." Nuclei of the nitrogen atoms were split when struck by these atomic bullets. Rutherford had caused the first artificial splitting of an atom. This is now called *atomic fission*. **Fission is the splitting of the nucleus of an atom into smaller parts.**

By the 1930's, machines were invented that acted as guns to fire "atomic bullets." These machines are used to accelerate electrons, protons, and alpha particles to speeds high enough to enter a nucleus and cause fission. These machines are better known as *particle accelerators* or *atom smashers*. Particle accelerators can fire atomic bullets at much greater speeds than those from natural radioactive sources. They are used to split nuclei of all types of atoms.

Nuclear physicists have found that the total mass of the particles from a split nucleus is slightly less than the mass of the nucleus before the split took place. They also found that very large amounts of energy are released when fission takes place.

Albert Einstein was the first to explain what happened to the mass destroyed during atomic fission. He pointed out that the destroyed mass had been changed into energy.

The energy obtained from splitting one atom is not very large. But if a large enough number of atoms are split very rapidly, a great amount of energy is released. Just before World War II, a way was found to split many atoms very rapidly. The material used to do this was a type of uranium called U-235. A device called a *nuclear reactor*, or *atomic pile*, was developed by scientists to split atoms of U-235 (Fig. 7-2). In the nuclear reactor *neutrons* were used as atomic bullets. When a speeding neutron hit a U-235 atom, this atom split and sent out more speeding "neutron bullets" that, in turn, continued to split nearby U-235 atoms. Every time a U-235 atom was split more neutron bullets were sent out and split other atoms, and so on. This continuous fission, or splitting, is called a *chain reaction*. **During a chain reaction a great amount of heat energy is produced.** This was the way that the first atomic bomb was exploded on July 6, 1945.

The heat energy produced by fission in a nuclear reactor can be used to produce the mechanical energy needed to run ships, submarines and electric power plants.

Even larger amounts of heat energy are possible with atomic fusion. **Fusion is a process in which nuclei of atoms combine**

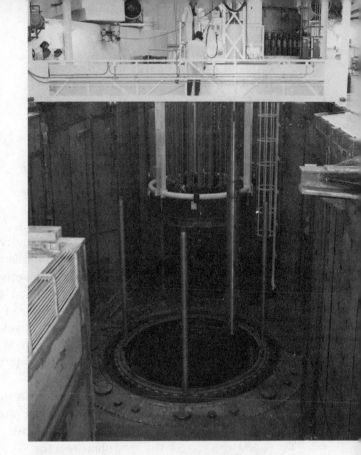

FIGURE 7-2 Fuel is being loaded into one of the Zion Units at the Commonwealth Edison nuclear plant in Illinois. Each unit is a nuclear reactor. (*Commonwealth Edison Company*)

to form larger ones. As in fission only a small amount of mass is destroyed, or changed into energy. Fusion occurs naturally when temperatures are high enough — millions of degrees. The sun and other stars have temperatures high enough for atomic fusion to take place. The sun contains a large amount of hydrogen. At the very, very high temperatures, hydrogen fusion can take place. The result is to form helium. Fusion of this kind produces enormous amounts of energy. The sun's light, heat, and other forms of radiant energy are the result of this kind of fusion. A hydrogen bomb is exploded by fusion.

Scientists are looking for ways to make use of fusion as a source of energy. A nuclear reactor using fusion would have advantages over reactors that use fission. A fusion reactor could produce more energy per amount of fuel used. A fusion reactor would also have less radioactive waste material to be disposed of. This is one of the big problems with fission reactors. A fusion reactor would probably use a type of hydrogen, which could easily be obtained from water in almost unlimited amounts.

OBJECTIVE 7
ACCOMPLISHED?
FIND OUT.

1. What is radioactivity?
2. What is an alpha ray? A beta ray? A gamma ray?
3. What is radioactivity? radioactive decay?
✳ 4. Suppose you had 40 grams of a certain radioactive substance with a half-life of 3 years. How much of the substance will remain after 3 years? After 6 years? After 9 years?
5. What is fission? What is fusion?
6. What happens to the mass that is destroyed during fission or fusion?
7. What is a chain reaction?
8. Give the main purpose of each of the following tools of the nuclear scientist: geiger counter, particle accelerator, nuclear reactor.

8 | ENERGY SOURCES — NOW AND TO COME

YOUR OBJECTIVE: To understand what energy is and how people have been using it for centuries; to know about some of the commonly used sources of energy; to see how the misuse of energy can be bad for the Earth's environment; to identify some new possible sources of energy and ways to conserve energy.

Everyone is becoming more familiar with the term *energy*. Almost every day we hear or read about an *energy crisis,* the need for more energy, or the search for new sources of energy. Energy enables cars, trains and airplanes to carry people great distances. It turns the machines that manufacture goods. It even enables us to move our own bodies.

Early people used muscle energy to do work. Later, animals were used to help people do work — oxen pulled plows and people rode upon the horse. The windmill and water wheel were later invented. These made use of the energy in the wind

FIGURE 8-1 Coal–stripping spoil piles in New Mexico — a desolate scene.

and moving water. For centuries they were used for doing such work as pumping water and grinding wheat. The invention of the steam engine in the eighteenth century and the gasoline engine in the next century provided two great energy sources. In fact, the industrial revolution, (mid 18th century to mid 19th century) would not have taken place without these devices. Many people think that the world's need for energy is second only to its need for food.

Where can the large amounts of energy needed in an industrialized world be obtained? **The United States gets most of its energy from the fossil fuels. These fuels include coal, petroleum, and natural gas.** All fossil fuels come from plants and animals that died millions of years ago and remained in the Earth. And they cannot be replaced when they are used up.

Coal was the most commonly used fossil fuel during the first part of the twentieth century. By the middle of the century, oil and natural gas had become more important than coal. Today, coal supplies about 20 percent of the total energy used in the United States. Coal is commonly used to supply the energy needed to turn

electrical generators. Coal is the most abundant fossil fuel in the United States. Even so it is estimated that the coal reserves can last less than a few hundred years if coal continues to be used as it is now used.

There are environmental problems when mining coal. Coal mines can leave great scars on the Earth's surface (FIG. 8-1). Many people oppose certain coal mining techniques for this reason. The burning of some coal can seriously pollute the air. Ways are being found to reduce these harmful effects on the environment.

At the present time, oil and natural gas supply about 75 percent of the energy used in the United States (FIG. 8-2). They are used to produce about 40 percent of the electric power. The known supplies of oil and natural gas are much less than that of coal. Experts tell us that the world supply of underground oil and gas could become very low within 25 years. And the supply in the United States could be used up before that time.

Scientists are looking for new ways to obtain and use the fossil fuels so there will be little bad effect on the environment. Large amounts of oil can be obtained

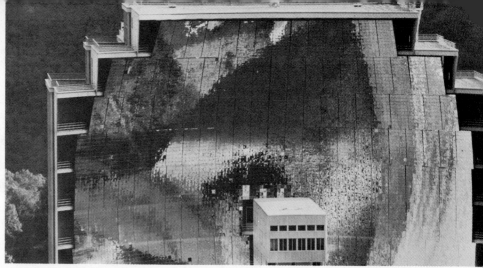

FIGURE 8-2 Oil rig. **FIGURE 8-3** This is the world's largest solar energy unit at Font Romeau in the French Pyrenees — it is six stories high. You can see the huge solar reflector. This device can melt a hole through one half inch of steel in less than a minute. *(Jeff Twine)*

from the oil *shales* that cover large areas of the western United States. To get at the oil, large areas must be stripped of the soil and vegetation covering the shale. The oil shale must then be mined and crushed. There are several different ways to remove the oil from the crushed shale.

Several ways are being developed for changing coal into a fuel that will be less harmful to the environment. A clean-burning gas or liquid fuel can be obtained from a very low grade of coal.

Scientists are looking for energy sources that can be used in place of the fossil fuels. Some of the sources being developed had been used in limited amounts even in the distant past. These include water power and wind.

> **Shale** Rock made from clay, mud or silt.

Electrical energy produced by the flow of water is called hydroelectricity. Water power is used to produce about 4 percent of our energy. The amount of hydroelectricity that could be produced in the United States is very limited. An almost unlimited amount of energy might be obtained from wind power. About one-half of the sun's energy that reaches the Earth is used to produce the motion of air. Large amounts of this energy is available without doing any harm to nature. **Scientists are now working on new ways to make use of the energy from the wind.** Windmills have been developed that can operate in winds of only a few miles per hour. Energy from a turning windmill would have to be stored so it can be used when the wind is not blowing. This energy might be stored in electrical batteries for later use.

One of the most important new sources of energy is nuclear energy. **A nuclear** power plant uses nuclear energy to produce electrical energy. Nuclear energy is released by nuclear fission, or the "splitting of atoms." More and more nuclear power plants are being built in the United States every year. These power plants will probably become a very important source of energy. Some people object to nuclear power plants because of the possibility of harm to the environment. The hot water produced in such a power plant can cause environmental damage if it is let out directly into lakes or streams. Getting rid of radioactive waste material from nuclear power plants is another environmental problem. There is a nuclear reaction called nuclear fusion that would be almost free of radioactive waste material. Nuclear fusion occurs when atomic nuclei come together, whereas nuclear fission is the splitting of a nucleus. Fusion takes place on the sun. There, under great amounts of heat — somewhere around 20 million degrees Celsius — fusion occurs when four nuclei of hydrogen atoms come together to form the nucleus of one helium atom. Because such a great amount of heat is needed for fusion, it is also called a *thermonuclear reaction*. It is very difficult to make enough heat to cause fusion.

Large amounts of energy that the Earth receives directly from the sun each day can be put to practical use. **Heat energy from the sun is called solar energy** (Fig. 8-3). The sun sends more energy to the Earth than we could ever want. The problem lies in finding an inexpensive way of using that energy.

Solar cells can take the sun's energy and change it to electricity. The electricity can then be used to do work. It might be used to operate a radio or even run an automobile. Solar cells have been used in outer space to supply the electricity in spacecraft. At the present time it is very costly to make large amounts of electricity in this way.

Large curved mirrors are used to catch the sun's energy. These mirrors reflect the energy into a small area. The concentrated energy can cause temperatures of thousands of degrees. Because of the high temperatures they produce, these devices are called *solar furnaces*.

Another source of energy is geothermal energy. **Geothermal energy is the heat energy that can be obtained from geysers and hot springs inside the earth.** Like the wind, the ocean currents might also turn generators capable of producing large amounts of electrical energy. The energy of ocean tides has already been turned into useful electrical energy in France.

With so many different energy sources, why do industrialized nations find a shortage of energy? For many years, the industrialized nations have relied almost entirely on the inexpensive fossil fuels for energy. There was no immediate need to develop other energy sources. The fossil fuels were used as if they would never run out. However concern for the environment is preventing the use of some fossil fuels for energy. Low-grade coal cannot be used because it causes air pollution. Concern for the environment also prevents certain kinds of mining from being used for some fossil fuels. If the industrialized nations are to survive, they must find more efficient ways to use the energy sources they have.

Aden and Marjorie Meinel of the University of Arizona, have another idea for using the sun's energy. They would like to see large areas in the southwestern U.S. covered with "solar farms." A solar farm would be made up of many long rows of lenses pointed toward the sun. The rows of lenses would direct the sun's energy on pipes located beneath them. This would cause the nitrogen gas in these pipes to become very hot. The heat in the nitrogen gas could be used to turn machinery that can produce electricity. The Meinels estimate that large amounts of energy can be generated in this way. A solar farm about one square mile in size could supply the energy needed by a city of about 60,000 people.

Energy must not be wasted needlessly. Homes must be insulated better to prevent heat loss in the winter. Cars that consume large amounts of gas cannot be used. Since much energy is used in manufacturing, manufactured goods must be used for longer periods of time. When the manufactured goods are finally thrown away, the materials in them must be "recycled." Even with an all-out effort to use energy wisely, new energy sources must be developed. We will not be able to rely entirely on the fossil fuels much longer. New and better energy sources, such as nuclear fusion, moving air and water, and the sun, must be found and developed.

Recycle Repeat the steps in a process.

Another possible means of energy conservation is geospace design, in which underground buildings are constructed. It has been found by calculation that there would be up to a 75 percent saving of energy used to heat earth-covered buildings as compared with above-ground buildings. Scientists are beginning to work on the possibility of building underground.

OBJECTIVE 8
ACCOMPLISHED?
FIND OUT.

1. What is a fossil fuel? What is the most abundant fossil fuel?
2. What is the commonly used source of energy in the United States?
3. What is meant by the term geothermal energy? by the term geospace?
4. According to experts, when will the underground oil and gas supply in the United States be used up? the coal supply?
5. Discuss several energy sources that might be used instead of "fossil fuel" energy. What problems must be solved to use these sources?
✳ 6. What are some reasons for the shortage of energy?
✳ 7. How might the energy shortage be overcome?

IN THIS UNIT YOU FOUND OUT

To the scientist, work is done when a force causes an object to move. Work equals force times distance. When work is done to lift an object, the force needed to move it is equal to the weight of the object. Power is defined as the amount of work done in a certain unit of time.

Energy is sometimes defined as the ability to do work. Energy is put into a machine when work is done on the machine. No machine can do more work than the amount of work done upon it.

The six simple machines are the lever, wheel and axle, pulley, inclined plane, wedge, and screw.

The mechanical advantage of a machine compares the resistance with the effort that moves it.

The efficiency of a machine is defined as the work put out by the machine (output) divided by the work put into the machine (input).

Work is a transfer of energy.

Energy that is active, or in the process of being transferred, is called kinetic energy. Energy that is stored is called potential energy.

The total energy can be found at any point in an object's motion by adding the potential and kinetic energies at that point. The total energy is always the same. Therefore the energy is said to be conserved.

Sound is a form of energy that is caused by the vibration of matter. Sound travels in longitudinal waves. A change in pitch because of the motion of the source of sound is called the Doppler effect.

The form of energy that causes the temperature of an object to change is called heat energy. Heat travels through solid materials by conduction. The movement of heat caused by the flow of atoms and molecules is called convection. Convection occurs in liquids and gases. The movement of heat energy in the form of invisible waves is called heat radiation.

Temperature is the measure of the "hotness" or "coldness" of a substance.

Electrical energy is called electricity. Static electricity is the momentary movement of electrons between unlike materials as a result of contact or friction.

Electric current occurs when electrons flow, usually through wires. A force, called an electromotive force (EMF), or voltage, is needed to make electrons flow.

A battery consists of one or more wet or dry electric cells. Electrons flow through a battery in one direction only — from negative to positive.

Electron flow in one coil of wire can be induced into a nearby coil of wire because of a magnetic field that builds up between the coils. This is called electromagnetic induction.

An electric generator changes mechanical energy into electrical energy. An electric motor changes electrical energy into mechanical energy.

An element that provides electrical resistance in a circuit is called a resistor. Ohm's law says that current depends directly on voltage and indirectly on resistance. Resistances can be connected in series or in parallel in a circuit.

Electronics deals with the flow of electrons through vacuum tubes or transistors. The cathode-ray tube is a special type of vacuum tube.

Radioactivity is the release of invisible rays from the nucleus of an atom. Fission is the splitting of the nucleus of an atom into smaller parts.

Fusion is a process in which nuclei of atoms combine to form larger ones.

The United States gets most of its energy from the fossil fuels. These fuels include coal, petroleum, and natural gas. Electrical energy produced by the flow of water is called hydroelectricity.

A nuclear power plant uses nuclear energy to produce electrical energy. Heat energy from the sun is called solar energy.

Geothermal energy is the heat energy obtained from geysers and hot springs inside the earth. Another possible means of energy conservation is geospace design, in which heat-saving underground buildings are constructed.

UNIT OBJECTIVES ACCOMPLISHED? FIND OUT.

Part A

Match the numbered phrases in the left-hand column with the lettered terms on the right.

1. The ability to do work.
2. Any device that makes work more convenient.
3. The transfer of energy.
4. Sometimes called "energy of motion."

a. Becquerel
b. Doppler effect
c. Einstein
d. energy
e. wedge

5. Sometimes called "stored energy."
6. The change in pitch because of the motion of a sound source.
7. A simple machine.
8. A measurement of the kinetic energy of the atoms in a substance.
9. The scientist who showed the relationship between current, voltage, and resistance.
10. The scientist who discovered radioactivity.

f. kinetic energy
g. machine
h. Ohm
i. potential energy
j. power
k. temperature
l. work

Part B Choose your answer carefully.

1. (a) Power (b) Work (c) Mechanical advantage (d) Frequency is equal to the product of force and distance.
2. A (a) foot-pound (b) horsepower (c) newton-meter (d) joule is not a unit used to measure work.
3. (a) Energy (b) Frequency (c) Temperature (d) Power is the amount of work done in a certain unit of time.
✵ 4. A certain lever has an effort arm of 8 meters and a resistance arm of 2 meters. Its mechanical advantage is (a) ¼ (b) 4 (c) 10 (d) 32.
5. Before any machine can do work, it must (a) be oiled (b) have energy put into it (c) have all friction removed (d) use a fossil fuel for energy.
6. A unit used for measuring energy is the (a) horsepower (b) watt (c) ohm (d) joule.
7. The kinetic energy of an object depends on its (a) mass and shape (b) shape and speed (c) mass and speed (d) speed and height.
8. As an object falls toward the ground its (a) potential energy remains the same (b) potential energy gets smaller (c) kinetic energy remains the same (d) kinetic energy gets smaller.

9. The range of hearing for a person is from about (a) 5 to 500 (b) 20 to 20,000   (c) 200 to 5,000   (d) 20,000 to 2 million hertz.

10. The frequency of the sound produced will increase if (a) a string is made tighter   (b) a string is made heavier   (c) a string is made longer   (d) an air column is made longer.

11. Sound travels fastest through (a) solids   (b) liquids   (c) gases (d) a vacuum.

12. (a) Sound   (b) Light   (c) Heat   (d) Temperature   is the energy that causes things to become warm.

13. The method by which heat travels through a vacuum is called (a) convection   (b) electron flow   (c) conduction   (d) radiation.

14. Zero degree on the Celsius temperature scale is the *same* temperature as (a) 0   (b) 32   (c) 100   (d) 212   degrees on the Fahrenheit scale.

15. Suppose you had 16 grams of a radioactive substance that had a half-life of one year. At the end of three years, (a) 2   (b) 3 (c) 4   (d) 8   grams of that substance would remain.

16. The splitting of a nucleus into smaller parts is called (a) radioactive dating   (b) fission   (c) fusion   (d) a chain reaction.

17. Mass is destroyed during fission and fusion. The mass changes into (a) a gas   (b) an electrical current   (c) matter   (d) energy.

18. The flow of electrons through a wire is called (a) current (b) power   (c) voltage   (d) resistance.

19. The voltage and frequency of electricity most used in a home is (a) 60 volts at 110 hertz   (b) 12 volts at 60 hertz   (c) 110 volts at 60 hertz   (d) 12 volts at 110 hertz.

20. A transistor acts most like a (a) vacuum tube   (b) battery (c) generator   (d) cathode-ray tube.

Part C   Think about and discuss these questions.

✻ 1. Suppose another student told you he or she had invented a machine that could increase force and distance at the same time. Comment on this.

✻ 2. What is meant by the phrase *energy is conserved?* Does there seem to be any exception to this?

3. Why will people be forced in the future to stop using some of the presently used energy sources? What will people do when these energy sources can no longer be used?

Index

Capillary action, 25
Capillary tubes, 24
Carbon, 51, 71, 73, 78, 79, 88, 106, 107, 108, 109,
  110, 118, 119, 121, 125, 126, 134
Carbon dioxide, 78, 79, 81, 85, 106, 107, 112, 119,
  126-127, 129
Carbon monoxide, 78, 79, 81, 83, 106, 107, 112, 118,
  126, 131-132, 134
Carbon-12, 62
Catalytic converters, 134, 135
Cathode-ray, 276
Cells, 122
Celsius, anders, 265
Celsius scale, 265
Centigrade, 265
Centimeter, 4
  cubic, 5
Centrifugal, force, 166-167, 182
Centripetal force, 166
Chain reaction, 279
Change
  chemical, 99-100, 101, 103
  physical, 99, 100, 103
Char, 112
Charcoal, 110, 118
Chemical bonds, 71-76
Chemical change, 99-100, 101, 103
Chemical energy, 255, 256, 258, 264
Chemical equations, 103, 104
Chemical formulas, 103, 74, 78, 88, 89, 103
Chemical properties, 63, 99, 113
Chemical reactions, 101, 102, 131, 134, 139, 140
Chemical symbols, 58-59, 61, 77, 103, 113
Chemistry, 51, 56, 79
Chemists, 53
Chile saltpeter, 125
Chlorine, 58, 64, 71, 78, 88, 91, 100, 138
Chromium, 116, 121
Citric acid, 87
Coagulant, 138
Coal, 105, 106, 108-110, 111-112, 132, 281, 282, 283
  acthracite, 108
  bituminous, 108, 109
Coal gasification process, 111
Cobalt, 117
Coefficients, 104
Cohesion, 25-26
Coke, 109-110, 112, 120
Collisions, 170, 172
  elastic, 170
Combustion, 106, 107, 109, 134
  complete, 107
  incomplete, 107, 108, 132
Compounds, 77-83, 87, 118, 122, 125, 127
Compression, 27-28
Condensation, 21
Conduction, 263
Conductors, 114, 263, 267, 272
Contraction, 38, 40
Convection, 41
Convection, 264
Converters, catalytic, 134, 135
Copper, 113, 114, 116, 117, 118-119, 121-122, 123
Copper iron sulfide, 123
Copper oxide, 118, 119
Copper sulfide, 119

Corrosion, 115
Covalent bonding, 71, 73, 78
Crest (wave), 225
  Crude oil, see Petroleum
Cryolite, 119
Crystal, 17
Crystalline structure, 17, 22
Cubic centimeter, 5
Curie, Marie, 54, 277
Curie, Pierre, 54
Current
  alternating, 271
  direct, 269
  electric, 267-275
Cyanide, 138
Cyclotron, 69, 70, 236

Dalton, John, 54
Deceleration, 151
Deformation, 29-30
Democritus, 53, 54
Density, 13-15, 28, 36
Deposits, 118, 124
Diffraction, 226
Direct current, 269
Displacement, 147, 151
Distillation, 86, 110
Doppler effect, 261-262
Double bond, 73
"Dry ice," 21, 127
Dynamite, 101, 125

Eclipse, 197
  lunar, 198
  solar, 197
Efficiency of machines, 250-251
Effort, 249
Effort arm, 249, 251
Einstein, Albert, 58, 59, 162, 234-236, 279
Einsteinium, 58
Elastic limit, 29
Elastic objects, 28-29
Electric current, 267-275
Electric field, 186
Electric force, 184-186
Electrical energy, 255, 256, 258, 264, 267-276, 282,
  283
Electricity, 267, 283
  static, 267
Electrodes, 268
Electrolysis, 119, 121
Electrolyte, 268
Electromagnetic induction, 270
Electromagnetic spectrum, 230-231
Electromagnetic wave, 230
Electromotive force, 268
Electron dot model, 73, 76
Electron flow, measurement of, 271
Electronics, 275
Electrons, 54-57, 60, 65-67, 71-76, 122, 185, 186,
  267, 268, 269-271
  valence, 76